3D PRINTERS FOR WOODWORKERS

A SHORT INTRODUCTION

Henry Doolittle

LINDEN PUBLISHING

Fresno, California

LINDEN PUBLISHING
The Woodworker's Library®

2006 South Mary Street, Fresno, California 93721
(559) 233-6633 / (800) 345-4447 / www.lindenpub.com

Book design by Andrea Reider
Cover design by Tanja Prokop, www.bookcoverworld.com

ISBN 978-1-933502-03-8

135798642

Printed in the United States of America
on acid-free paper.

Library of Congress Cataloging-in-Publication Data

Names: Doolittle, Henry, author.
Title: 3D printers for woodworkers : a short introduction / Henry Doolittle.
Description: Fresno, California : Linden Publishing, [2020] | Includes index.
Identifiers: LCCN 2020042551 | ISBN 9781933502038 (paperback) | ISBN 9781610353762 (epub)
Subjects: LCSH: Three-dimensional printing. | Woodwork--Equipment and supplies.
Classification: LCC TS171.95 .D66 2020 | DDC 621.9/88--dc23
LC record available at https://lccn.loc.gov/2020042551

Table of Contents

Foreword

3D Printers for Woodworkers offers a compressive look at the history, development, and usage of 3D printers. To understand a tool, it helps to understand its history: how it came about, how it was used, and how it evolved to its use today.

Once the history lesson is done, Henry Doolittle dives into the nitty-gritty of 3D printing. What types of printers are there? Which 3D printer is best for particular applications? What filaments do you need for your particular project? He answers all these questions along with discussing where to place your printer and how to calibrate the print head. Then Henry gives you projects to test your printer's capabilities.

He shows successful prints and even more importantly shows you prints that go wrong. Then he explains in detail how to avoid the issue or fix it.

History and education are great, but a book on 3D printing had better share some useful and fun projects, and Henry does not disappoint. He provides well over a dozen projects for the woodworker. From a center finder to a pocket screw jig to a corner jig, Henry lays out the steps and tips needed to become a competent 3D craftsman.

If you want to test out this new technology, I highly recommend you purchase a copy of *3D Printers for Woodworkers* as your introduction to the world of 3D printing.

—Tim Yoder, nationally awarded television producer, photographer, editor, and longtime woodturner

Chapter 1

A History of 3D Printing

Major manufactures, such as Volkswagen and General Electric, have recently announced their intent to use 3D printers to mass produce parts.

—George Takei, actor

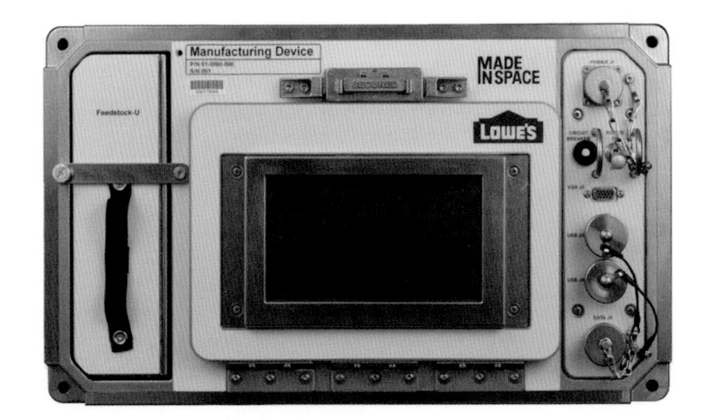

A 3D printer built by Made In Space for use on the international Space Station. *Made In Space*

In a 2013 study on 3D printers, "Life-Cycle Economic Analysis of Distributed Manufacturing with Open-Source 3-D Printers,"[1] the authors demonstrated that the average household could save between $300 and $2,000 a year by printing items that they would normally purchase. In the study a college professor had his students look at the website Thingiverse and find items that homeowners would be buying on an annual basis at Home Depot. The cost savings were calculated based on the cost of the parts if purchased at Home Depot minus the cost to 3D print the same parts at home.

There are a number of websites offering 3D print files (known as STL files), including Thingiverse and MyMiniFactory. New print files are being generated daily, ranging from cosplay costumes to face shields for COVID-19. 3D printers are coming down in price and the quality of prints is increasing. Not long from now you will be able to go online at Home Depot or some similar website, pick out the tool or part needed, pay a nominal fee, download the print file for the part, and print it at home. In a couple of hours, you will have the part in hand, without leaving your shop.

The history of 3D printers is the history of computers. Without cheap computers 3D printers would never have come into existence. The two are inseparable. It takes a computer to generate the fine movements of a 3D printer's print head. And to get 3D printing into the average home requires cheap computers. With faster, cheaper computers and controller boards, 3D printers will get better and easier to use.

The International Space Station (ISS) is about as far off the grid as you can get. When there's a problem, a quick run to the local hardware store isn't an option. If crew members don't have the tool they need, it is sent up on the next ship, which can be several months away. When they lose a tool, it's lost. It's not like that 10 mm socket or tape measure you just dropped on the floor and bent over to pick

1 www.sciencedirect.com/science/article/abs/pii/S0957415813001153

up. When astronauts drop a tool, they have no choice but to watch it drift away. At some point in time it is going to drop back to Earth and end up in someone's backyard.

NASA had a problem. How do you ensure that the ISS has the tools and parts needed to keep the station operational? In spite of its budget, the ISS does not have a tool room with an unlimited number of tools. The ISS can carry only so many 10-mm sockets. What size patches do you need? How many plumbing fittings do you take with you? Do you have the parts needed to fix the Wolowitz waste disposal system?

Having the parts on hand for every conceivable repair was not an option. They needed a method of producing necessary tools and parts when they needed them. They needed to be able to create and repair with the materials on the space station.

Enter 3D printing. Under contract with NASA, a company called Made In Space built a 3D printer for use in space. The printer was a FDM (fused deposition modeling) design made to work in a zero-gravity environment. In March of 2016, their 3D printer, called the AMF (additive manufacturing facility), was sent to the ISS. It has since printed over one hundred mission critical parts. NASA engineers have estimated that 30 percent of the parts in the ISS can be printed by a 3D printer. The only materials that need to be replenished for these printers are the rolls of filament.

The AMF is small, about the size of a toaster oven, with a build volume of 14 × 10 × 10 cm (5.5 × 4 × 4 inches). It is built to NASA standards, but it uses the same types of parts and works the same as all other FDM printers. And it uses the same filaments available for use with current hobby FDM printers. In fact, it is even possible to have a part printed on the ISS and have it delivered to Earth with the next space launch. Sure, the cost starts at $12,000, but you will have a part made in space! Made In Space is currently working on a 3D metal printer for use on future space missions.

Design engineers have dreamed of instant prototyping as long as there have been engineers. The idea that you could design, make, and test a part before sending it to the factory has been around a long time. Engineers frequently had a small team of machinists and model makers to test ideas. Even with these dedicated teams it could still take months to make a part for testing. Using 3D printers for rapid prototyping can be done in days, greatly reducing the time to market for new products.

Three-dimensional printers have existed in science fiction dating back to the 1950s. Popularly known as replicators, these marvelous devices have been found on such TV shows and movies as *The Jetsons*, *Star Trek*, *Stargate*, and *Harry Potter*. The idea of instant gratification was not lost on science fiction writers. It also solved the problem of creating technology without needing to explain it.

We remember George Jetson. He lived in an apartment in the sky, went to work in a flying car, and worked a three-day workweek pushing a button. Jane worked hard to put dinner on the table. She had to push a button and dinner appeared, including dishes and silverware.

The first fictional use of 3D printers was for creating food, an idea that was not lost on Gene Roddenberry, the creator of *Star Trek*. *Star Trek* in all of its iterations included 3D printers/replicators in various designs and for various purposes. More of the printers appeared in medical scenes or contexts than in any other.

In *Stargate* the replicators became self-reproducing. They started out as a toy for the inventor's daughter. The toy got out of control and became the villain of choice. The replicators would join to form bigger replicators and were able to take local materials and build new replicators. When you add in artificial intelligence, the replicators became a formidable opponent.

No one can forget the 3D printer used to re-create Leeloo Dallas in the movie *The Fifth Element*. It was probably one of the more realistic 3D printers used in the movies and rather predictive, given the uses for 3D printers in the medical industry. The printer built Leeloo using additive methods. The same methods are being used today to print bones and organs. A 3D-printed kidney has been made and is being tested for use in humans.

Commercial 3D printers are available for making pancakes to order and to the desired shape. You input the desired shape, and the printer places the pancake mix on the hot plate. When the pancake is done, you flip it and eat. There are also commercial 3D chocolate printers and candy printers. A pizza printer would not be out of hand, and in the future a printer that produced custom casseroles or side dishes would not be inconceivable.

Small homes are being printed using concrete. In India there is a need for housing in a part of the country that sees frequent cyclones. Builders are working on printing a 320-square-foot home in three days and make it ready for habitation within a week. India is looking at printing the first such home in 2021, if not earlier. The size, shape, and materials used make the homes cyclone proof.

Programmable subtractive machining was available back in the 1700s, as was documented in Charles Plumier's *L'art de tourner en perfection* (1749). Programming was performed using various gears and cams to achieve the desired results. Known as a rose engine, or guilloché lathe, these machines were used to provide decoration for wood, metal, or ceramic items. Fine decoration could be generated by changing cams on the lathe. At least one was used by the House of Fabergé, a jewelry firm founded in 1842 in Saint Petersburg, Russia, in the creation of the Fabergé eggs.

As the technology improved, the machines became more complex, providing more options for engraving. Gearing also allowed the creation of metal lathes and milling machines that provided a level of programing.

In the nineteenth century, Charles Babbage worked on gear-driven programmable computers. Although Babbage was never able

A Fabergé egg with Guilloché design cut with an ornamental lathe.
© *Can Stock Photo / TpaBMa.*

to achieve a programmable computer, primarily due to the inaccuracy of the machines, he made major advances in machining and gear operations.

In World War II, the Enigma machine generated coded messages for use by the German military. The machine used an electromechanical rotor system to encode the messages, which were considered unbreakable. Each day a key was entered into the machine to set the code for the day. It was necessary to know the key in order to decipher the message.

Alan Turing, an English mathematician and computer scientist, developed an electromechanical machine, the Turing machine, which was able to calculate the key needed to decipher the messages. Alan Turing is considered the father of theoretical computer science. The Turing machine was one of the first general-purpose computers.

John von Neumann is considered the father of the modern computer. He was one of the mathematicians who worked on the Manhattan Project. In the late 1940s he convinced Princeton University to fund a project that produced the first all-electric, programmable computer using current computer architecture.

In the 1960s computers were used to control motors used to drive metal- and wood-cutting machines. These machines were built on the concept of an *x*/*y* plane with a *z* axis moving the cutting head in and out of the target material. This is called the Cartesian system.

Although the idea of rapid prototyping, or additive manufacturing, has been around for a long time, the technology wasn't available until the mid-1980s. The limiting factor was the computing power required to calculate the moves in three directions (the *x*, *y*, and *z* planes). Such computing power wasn't available until the 1960s, and was not affordable until the late 1980s.

To run a CNC (computer numerical control) machine, computing power is necessary. In the late 1970s, the computers needed to run a CNC were big and expensive. Then came Intel and microprocessors. The Intel 4004 microprocessor was developed for use in a calculator but was designed as a fully functioning computer. The Intel 4004 started a new industry and moved us one step closer to the production of 3D printers.

By the mid-1980s, computers made by IBM, Apple, Commodore, and others were available with the computing power and the process connections capable of operating a 3D printer. Between 1985 and 1990, four different people independently developed fully functional 3D printers. The computing power of personal computers provided the ability to generate the tool paths needed to print a model and control the stepper motors that control the printer. These computers were what allowed three of the four engineers to develop three different primary printer designs. This resulted in the companies 3D

An early Intel microprocessor. © *Can Stock Photo / Romanchuck.*

Systems Corporation, DTM Inc, and Stratasys Inc. being founded. These companies went on to place commercial printers into the marketplace by 1990. A French engineer also developed a 3D printer, but never moved to take the product to market.

In 2005 Dr. Adrian Bowyer, professor at the University of Bath, founded RepRap (*rep*licating *rap*id prototype). In 2002 Dr. Bowyer had received two 3D printers for use at the university. The cheapest of these printers was $50,000. Looking at the printers, Dr. Bowyer decided that he could use them to print parts needed to make a new printer. The new printer then could print the parts needed for another printer—a self-replicating printer. RepRap was thus designed and built as an open-source 3D printer able to reproduce parts necessary to build new printers. The design of RepRap was released under a general public license. The philosophy behind RepRap is to spread cheap 3D printing. To get a RepRap, you would find someone who owned a printer. They would provide you with a bag of printed parts, a bill of material, and instructions. When your printer was finished you would print out the parts for another printer and pass them along to someone else. Many current printer designs are based on the original RepRap design.

In 2008 the first prosthetic limb was made using 3D printing technology. 3D printing has driven the cost of prosthetic arms down to where home enthusiasts are now making fully functional prosthetic arms for children. A prosthetic arm that would normally cost $80,000 to $100,000 can now be printed at home for under $5,000.

By 2009 the first patents for 3D printing had expired, allowing more competition in the industry. The combination of expiring patent and the RepRap design becoming available opened the market for small companies to sell inexpensive 3D printers. And the rest, as they say, is history.

Over four hundred 3D printers are now available, and new ones are appearing every year. They cost from fifty dollars to hundreds of thousands of dollars. Companies are selling parts to make your own custom 3D printer. Printers come in sizes that range from a couple of inches to big enough to print bridges and homes. New printing methods are being designed, and new filaments are coming out all the time.

3D printers are being developed to solve problems that can't be solved using traditional methods. A metal bridge is created in the Netherlands, and a plastic bridge is made in Singapore. Small homes are being printed in India that are able to withstand the harsh weather seen on the India coast.

The future of manufacturing is in 3D printing and additive manufacturing.

Chapter 2

Basics of 3D Printing

> With 3D printing, complexity is free. The printer doesn't care if it makes the most rudimentary shape or the most complex shape, and that is completely turning design and manufacturing on its head as we know it.
>
> —Avi Reichental, CEO, 3D Systems

Imagine you want to make a toy for your new daughter. You take a glue gun to draw a picture of a frog using a low-temperature plastic. You start drawing the frog using the material you push out of the glue gun. It works but not very accurately. You decide to use a mechanism like the inside of an Etch A Sketch to control how you lay out the plastic. You use two knobs to control the movement of the glue gun. When you complete one layer, you raise the glue gun and start the next layer. Next you decide that you can add stepper motors to the knobs and thereby move (step) the motors to control distance and direction. Add a controller and you can tell the motors to move x 10 steps (along the x axis) and y 5 steps (along the y axis). This is how S. Scott Crump developed the first FDM printer.

In the world of computer-aided manufacturing (CAM), you have both additive manufacturing and subtractive manufacturing. We are used to and expect subtractive manufacturing: the use of saws, mills, planners, lathes, and other tools to remove stock to get to the final size and shape. We cut, grind, sand, and remove material as needed.

If you make cabinets, you start with a sheet of plywood. You cut it into smaller panels using saws. You gradually remove material and reduce the size of the original panel, taking the smaller panels to the next step, and so on.

CNC routers, for example, use a process of subtractive manufacturing. You start with a block of material and remove that which is not needed. With a CNC router you place the sheet of plywood on the bed and provide the instructions to the router. The router cuts the individual panels, drills the holes, and sets them up for sanding and glue-up. The CNC router is a tool just like any other tool in the shop. The difference is the level of automation and the amount of upfront effort required.

The CNC router requires CAD (computer aided design) software, to set up the layout of the cuts and holes. And computer aided manufacturing (CAM) software provides the G-code (G-code is a programing language that tells the CNC mill what to do) needed to direct the movement of the router's head. The same is true for 3D printers.

All 3D printers use a process called additive manufacturing, as opposed to the process of subtractive manufacturing. The normal subtractive process of building a wooden bowl is to take a block of wood and remove the material that isn't needed. A 3D printer works by *adding* material as needed to create the bowl, not unlike the process of segmented turning, whereby the bowl is built one layer at a time (an additive process) before finally turning away the edges to form a final shape (a subtractive process).

A CNC mill operates on three axes, *x*, *y*, and *z*, a Cartesian coordinate system. In the Cartesian system, any point can be identified based on the three coordinates or the distance from a specified point. The G-code will tell the CNC where to move the head. The code uses a simple syntax in the style of G1 *x*100 *y*100 (G=go, 1=slow speed, *x* direction 100 steps, *y* direction 100 steps).

Imagine building a 3D topographic map of Mount Rainier in Washington State. The normal process, called contour modeling, would be to take several topographic maps of Mount Rainier and glue them on to sheets of either plywood or cardboard. The maps are then cut out following individual contour lines at regular elevations. The sections are stacked and glued in order, on a table, to form a 3D model of the mountain.

This process is used at many national parks to show items of interest and elevations. Architects use the method to show the land

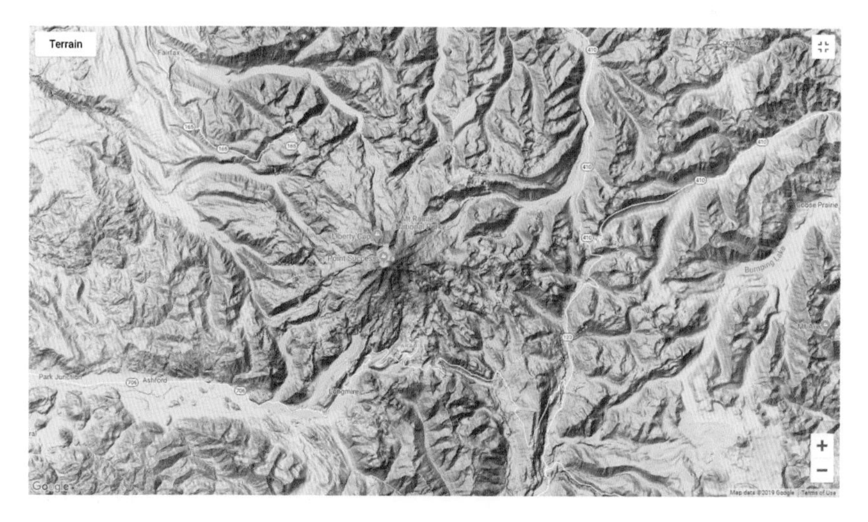

Topography of Mount Rainier. *Terrain2STL (Thatcher Chamberlin)/Google Maps*

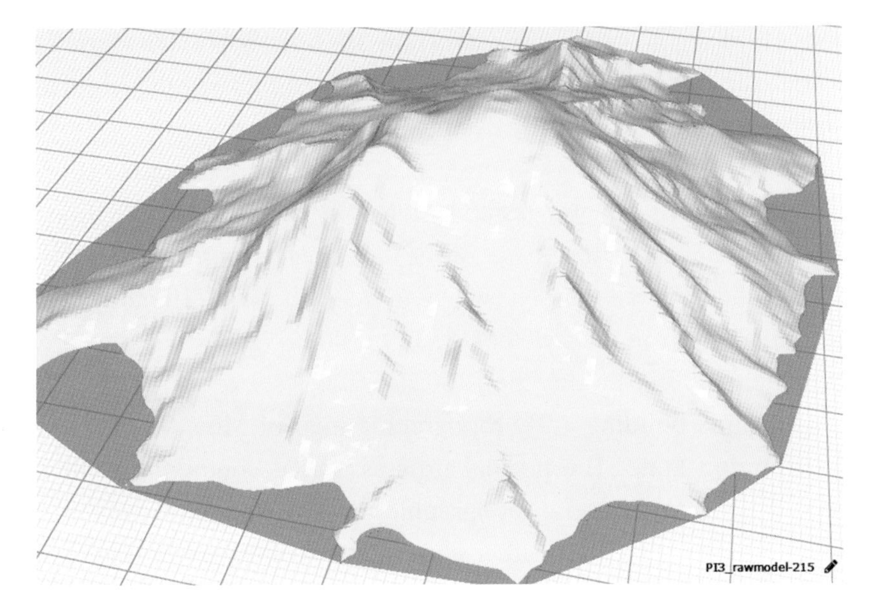

Mount Rainier model

around a building site. Model makers use similar methods to create models of parks and resorts.

Depending on the size of the model and the precision of the craftsman, the model can be a very accurate representation of a landscape, showing the terrain built one layer at a time. However, these models can be very time consuming to construct, and very expensive if accuracy is called for.

If we take the data for Mount Rainier and run it through a CAD/ CAM program, we can generate the same model using a CNC router. The router will cut a block of wood, on a plane, one layer at a time. It removes the necessary material for that layer, then it moves down to the next layer and starts cutting again. The router will continue cutting down the block of wood until what remains is a model of Mount Rainier. How fine the representation is depends on the thickness of the layers. The layer lines can be left in place, or they can be removed, leaving a smooth contour line. The CNC will perform the same task

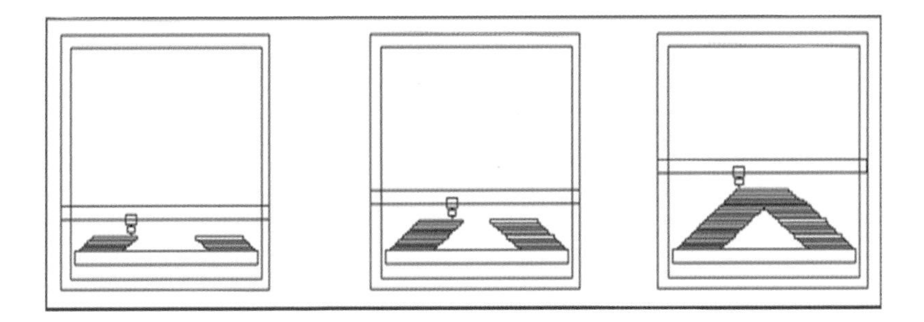

3D printing, layer by layer

as the model maker but much faster and with greater accuracy. The craftsman still has an important role to play as he goes from working with a coping saw to programming a design into the computer and setting up the CNC router.

The 3D printer works like the CNC router but in reverse, by adding material. Like the craftsman-built model, the printer will lay down layers of material to represent the contour lines. It will lay down one layer at a time on an *x* and *y* axis and then move up to the next layer, on the *z* axis, until the model is complete. Each layer is defined by the same data as with the CNC. The operator decides the accuracy of the print and the layer height. A model can be hollow or filled. You can design a model with an internal structure and place parts inside the model as it is built. You can have holes inside a printed part for captive nuts by pausing on a layer and installing the nuts. You then restart the printer to finish the part with the captive nut.

Because a 3D printer works by adding material layer by layer, designs can be made that can't be manufactured using any other method. The Veritas P1169 3D-Printed Plumb Bob is a design that can only be made through additive manufacturing. Veritas made the plumb bob to show what could be accomplished using the new technology.

A 3D printing of Mount Rainer's topography

There are very few industries that are not currently using some form of additive manufacturing. Dental offices have started printing crowns in the office, allowing a tooth to be replaced in one sitting instead of two weeks.

Four bridges have been constructed using additive manufacturing technologies: one in the Netherlands, one in Spain, and two in China. The one in the Netherlands was built in place using robots to splatter-weld the bridge from both ends, meeting in the middle. The bridge in Spain was built using microfiber-reinforced concrete. And the bridges in China were printed using ABS plastic and a more traditional but large (very large) FDM printer.

Children's prostheses are currently a big use for 3D printers. Prostheses are expensive, and kids continually grow out of them. Several organizations are providing designs and information to owners of 3D printers, allowing them to make hands and arms for kids. One of these organizations is e-Nable.

A Veritas 3D-printed plumb bob. *Lee Valley Tools Ltd.*

Jewelers have adopted 3D printing, especially with the use of stereolithography (SLA) printers. The resins used in SLA printers produce highly detailed prints that when placed in a furnace will burn without leaving ash. Prints from SLA printers work very well for lost-wax casting.

Dave Kindig and his team at custom-car shop Kindig-It Design use a 3D printer to create custom trim for their cars. The 3D printer gives their shop the ability to generate unique designs and builds. To create a special look, Dave and his team are able to use materials that are not typically found in cars.

Like CNC routers, 3D printers can make the craftsman's job easier. More tools and jigs can be made from a combination of metal and plastic. Most of the plastic parts can be printed. In the future you will be able to go online for the tool or jig you need, buy it, download it, and print it. Tools like speed squares, pocket screw jigs, feather boards, and clamps can all be printed on a 3D printer and made available to your shop as needed.

Chapter 3

Types of 3D Printers

> It would appear that we have reached the limits of what it is possible to achieve with computer technology, although one should be careful with such statements, as they tend to sound pretty silly in five years.
>
> —John von Neumann

The same can be said about 3D printers. Every year they are increasing in usefulness and quality. New designs and new filaments are available. Ten years ago there were no 3D printers for home use. Five years ago the quality and the filaments were minimal. Any comments we make today about 3D printers will sound pretty silly five years from now.

If you talk with most 3D printer manufacturers, you will hear their plans for making 3D printers common household appliances. Your printer could be a small box sitting in the corner of the garage. Let's say some part breaks in your house. You go to the garage, pull up the menu, find the part you need, and push the button. It's faster than Amazon, and you don't need to leave home. If you need tools you can go online, buy the print files, and print them out on your printer.

We are still several years away from the 3D printer being a home appliance, but we are getting there. 3D printers have already entered the small- and moderate-size business shop, manufacturing custom parts or parts that are no longer available.

Three basic types of 3D printer are available to the user: selective laser sintering (SLS), stereolithography (SLA), and fused deposition modeling (FDM). Each has different capabilities and limitations. Each functions better for some uses than for others. SLA printers work best for artwork and making models for casting. Because of the high cost of the equipment and the complex process involved with SLA printers, SLS printers work best for rapid prototyping. And FDM printers work best for creating working parts in a small manufacturing environment. For most woodworkers, the FDM printer is the best option.

Selective Laser Sintering (SLS)

Dr. Carl Deckard developed and patented selective laser sintering in 1986. He worked as a summer intern at a machine shop while going to Texas A&M. Although the shop had all the latest in milling machines, he was amazed at the amount of work that went into creating castings.

The process of metal casting involves pouring molten metal into a mold made of sand, metal, or ceramic. The resultant part can then be machined to the required tolerance.

Deckard spent the next several years developing the process for sintering metal powder to create machinable parts. (Sintering is a process where a metal powder is heated to the point where the metal fuses together.) During this period, he earned his PhD in engineering and started a company.

SLS printers can be fabricated to use a large range of materials. These materials are primarily metals such as steel and titanium, but almost any metal can be used, with the right printer. The SLS printer gets rid of the need for casting by sintering the metal into the required shape. This capability eliminates several steps in the process.

Depending on the need, the material can be partially or completely melted. The parts can achieve 100% density and, like particle metallurgy, they have none of the flaws found in cast parts. The layers of metal powder are on the order of 0.1 mm, so parts with very fine detail can be printed.

An SLS printer has two chambers, the powder delivery chamber (or powder chamber) and the fabrication (or print) chamber. Both chambers have plungers. The process starts by filling the powder chamber. The printer heats up to a temperature just below the melting temperature of the powder. When the powder chamber is full, the plunger rises 0.1 mm, the print chamber plunger goes down

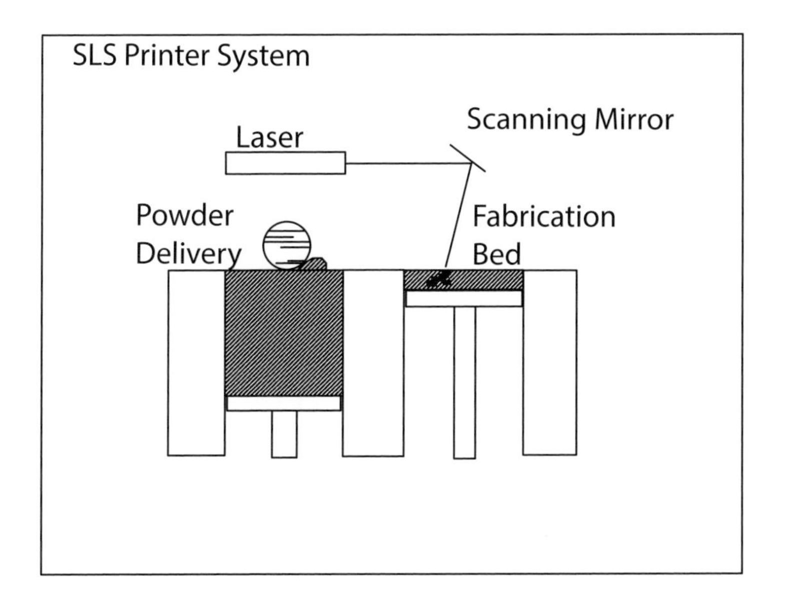

0.1mm, and a paddle moves across the top to move the powder to the print chamber. The laser writes the first layer on the top of the powder in the print chamber. The print chamber lowers, the powder chamber rises, and the process repeats for each layer until the part or parts are printed.

The parts are removed from the printer and taken to a sieving station where they are separated from the powder. The parts are then taken to the sandblaster for final cleaning, and the powder is run through the sieve to recover unused powder. The retrieved powder is taken to a blender to be mixed with new powder for the next part. Up to 50% of the powder can be reused. The process doesn't require a clean room, but must still meet lab-room-clean standards.

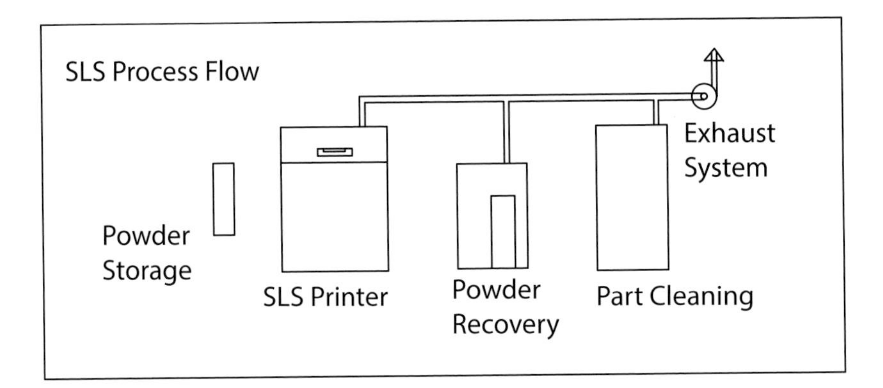

Low-end SLS machines are designed to work with nylon, specifically nylon 6 and 6a. The print volume is small at around 4 inches cubed (about the same volume as the AMF printer on the ISS). These small machines are aimed at engineering prototyping only.

However, the metal printing SLS machines are useful in all aspects of manufacturing. They are the most efficient method for creating one-off pieces or replacement parts for out-of-date equipment.

The disadvantages of SLS printers are as follows: They are expensive. They are material specific, meaning you need to get a different printer if you want to start printing with a different material. The materials used in SLS printers are expensive and messy. SLS printers need to be maintained in a controlled environment and used by a trained operator. SLS printers are becoming cheaper and more user friendly, but they still are not for the casual user.

Stereolithography (SLA)

Dr. Hideo Kodama devised the first SLA process in Japan in 1980. However, Kodama was unable to fully develop the process due to the high cost of computing at that time. Personal computers were just coming into use, the TI 99 and the Apple 2 being the main computers available. The IBM PC was not available until 1981.

By 1984 home computing power had increased to the point that Charles W. Hull could develop and patent the stereolithography printing process. Hull is an engineer who got tired of the long delays in prototyping. He came up with the idea for stereolithography and offered it up to his company, which rejected the idea. So he spent the next couple of years developing the process in his garage. Hull started a company called 3D Systems and introduced the first commercial 3D printer in 1987.

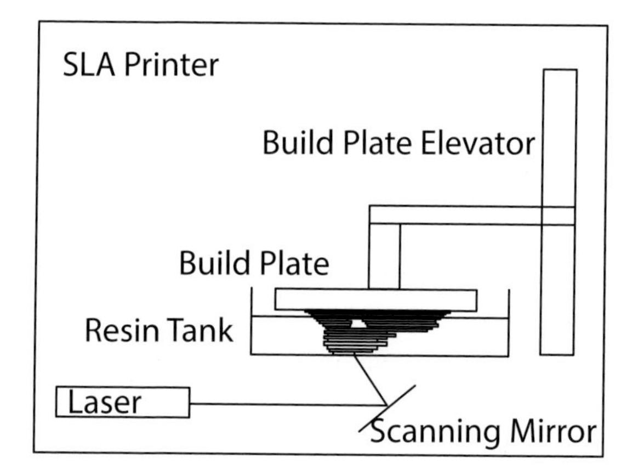

SLA printers use a light-activated resin. The resins are printer specific and dependent on the light source. Light sources include lasers, high-intensity lamps, and LCD (liquid-crystal display) screens. LCD screens are becoming the light source of choice for SLA printers. The resin is placed in a glass tray, and a metal plate is lowered into the tray to within 0.1 mm of the bottom. The light source is activated to draw the first layer, activating the resin. The plate is lifted and the next layer is created. The process continues until the part is completed. The plate is then removed and the part is washed off and hardened under a UV lamp.

SLA printers are available starting at $500, and a full system could cost over $5,000. SLA printers can provide a very fine print that works well in making jewelry. The parts also work well for casting. The parts are brittle, like nonreinforced epoxy. The main problem with SLA printers is the mess. The resin reacts to light and must be kept in a UV-free environment. When removed from the vat, the parts are wet and need to be handled with gloves until fully cured.

Fused Deposition Modeling (FDM)

S. Scott Crump developed the first FDM printer in the late 1980s. He wanted to make a toy frog for his daughter, so he started modeling with machinable wax (a mixture of wax and plastic) and a hot glue gun. He decided to try to automate the process and built a system of stepper motors to drive the glue gun. In 1990 he and his wife started Stratasys with the intent of producing and selling FDM printers. His first printer, 3D Modeler, was made available in 1992.

FDM printers are the cheapest and most popular. They range from a couple of hundred dollars to tens of thousands of dollars, depending on design. At the low end of the scale are FDM printers that have a small build plate and will only print with PLA filaments (see chapter 7 for full details on 3D printing filaments). High-end

FDM printers can print 8-foot-tall models or can print using multiple different materials in the same model to create working prototypes.

Hundreds of manufacturers are producing FDM printers, but not all of them are good. They are the easiest to make at home with available materials since the design is rather basic. Current software can be customized to work with any printer design.

There are two basic designs of FDM printers, Cartesian and delta. A Cartesian printer is named after the dimensional coordinate system. The G-code commands provided to the printer guide the head in the *x*, *y*, or *z* direction as specified. Most of the FDM printers on the market are of the Cartesian design.

The delta printer works over a circular bed with three arms. The arms move up and down to control the position of the print head.

Other designs are being developed that may be available in the future, but for now we have only these two choices, or variations of them. Both designs accomplish the same job using similar methods. The delta printers can usually provide taller prints, whereas the Cartesian design operates with a larger footprint.

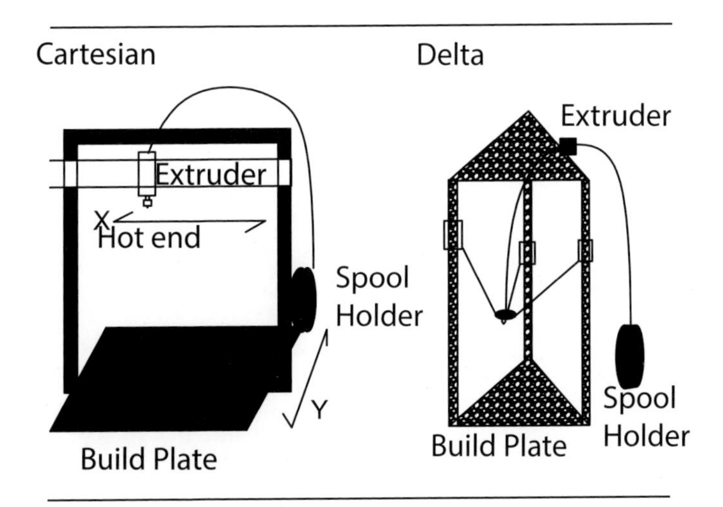

Multi-material Printers

Unlike SLA and SLS printers, the FDM printers can print using different colors and materials in the same print. Multi-material printers provide options. For example, you can print tools with writing on the side. Specific size information can be printed on the part using contrasting colors. And because parts can be printed using different materials, things such as vacuum parts can be printed from both PETG, for strength, and TPU, which provides a soft surface for sealing.

An additional material to consider using is a soluble filament. Soluble filaments, such as PolySupport, work for temporarily supporting internal structures (such as overhands) in the print where support filament can't be removed using other means. Soluble filaments are easy to remove.

One method of creating a dual-material printer is to make a custom hot end (the working part of the printer) that combines two hot ends and two filament extruders. The printer switches between hot ends depending on the material being printed. Most dual-material printers use this method, including Ultimaker and Flashforge. Hot ends are available for home-built printers that can use three or four materials at the same time.

Another method is to use one hot end and change filaments as needed outside the printer. This is the method used by devices like the Mosaic Palette and the Prusa MMU. Mosaic Palette can switch between four different filaments. Palette uses special software to calculate the length needed for each filament, in accordance with the G-code for the part. When the correct length of filament has been used, Palette cuts the filament in use and splices on the next filament. This process continues until the print is complete.

Prusa MMU follows a different method. It can use up to five filaments at the same time. It uses the same slicer software and is

A Prusa MK3 printer with a multi-material model in production

connected to the printer CPU. When the G-code calls for a different filament, the printer ejects the current filament and the MMU inserts the next filament.

For printers that use dual hot ends, the two filaments can be drastically different. For other multi-material printers, the hot end maintains a set temperature, so the filaments need to have similar temperature profiles.

A 3D print from a multi-material printer

Chapter 4

Definitions

> If you tell me precisely what it is a machine cannot do, then I can always make a machine which will do just that.
>
> —John von Neumann

Build Plate

The build plate is the plate that the printer prints on. The build surface needs to be completely flat and level to the *x* and *y* axes or the print won't work. The build plate rests upon the print bed. Depending on the printer, the print bed may have clips or magnets to hold the build plate in place.

A popular bed material is glass, but manufacturers are moving over to a removable, flexible build plate because of certain issues with glass. The glass plate is stable but sometimes needs tape, glues, or other materials to get proper adhesion to work. Also, glass plates are not easily removable, requiring the parts to be removed in place.

A newer type of build plate is a sheet of spring steel on a magnetic base. The spring steel is coated with a plastic

Various removable build plates

cover that provides greater adhesion with the print plastic. The plastic cover of choice is PEI (polyetherimide). When heated it provides good adhesion with the printing filaments. Other plastic sheets are available from third-party companies.

All of the build plates need to be leveled in order to provide a proper print. Different printers use different methods for bed leveling. One of the cheaper methods is to provide screws in the four corners of the bed. The hot end is placed on the corner and the corner is raised or lowered until it is just touching the hot end. This is repeated on all four corners until the bed is level. This process needs to be repeated every dozen prints to ensure that the bed is level.

The more current method is to use a probe to examine the bed each time the printer is used. The bed is probed in multiple locations in a grid pattern to provide a mesh map that is used to adjust the *z* axis to compensate for out-of-level conditions. Some printers can perform self-leveling, and will check the level of the bed every time a print is run.

Another bed option is a heated bed. The heated bed provides greater adhesion in printing. If the only filament being used is PLA, a heated bed is not needed. For ABS, the bed needs to be able to get to 125°C and then hold that temperature in order to reach proper adhesion and minimize warpage.

Enclosure

An enclosure is not necessary for a 3D printer, but it does help maintain the project at a constant temperature for the duration of the print. The constant temperature provides for better adhesion between layers and minimizes warpage. If the printer doesn't come with an enclosure, then one can be built or purchased. A cardboard box will work just fine as a temporary enclosure. Sturdier enclosures can be made from plastic or wood, and some have even been made out of IKEA LACK tables. Even an old photo box can be used. Also available are commercial enclosures using T-slot aluminum and plexiglass. There are companies that make enclosures for specific printers. All these options work, but they are not necessary for all printers.

The enclosure is not intended to be airtight. With the heat coming from the bed and the hot end, it is possible to overheat the electronics. That's why you don't want to hold in the heat, you just want to keep drafts from the hot end. The enclosure is there to keep breezes from cooling the plastic too fast.

Heated enclosures have their advantages. Maintaining an enclosure temperature above ambient can keep the plastic warmer, allowing greater adhesion between layers and minimizing warpage. A heated enclosure needs to be controlled, and temperatures greater than 50°C can damage electronics and prevent proper cooling at the print head. Without proper cooling, the filament in the hot end can degrade and plug the hot end. For this reason, there are very few FDM printers with heated enclosures.

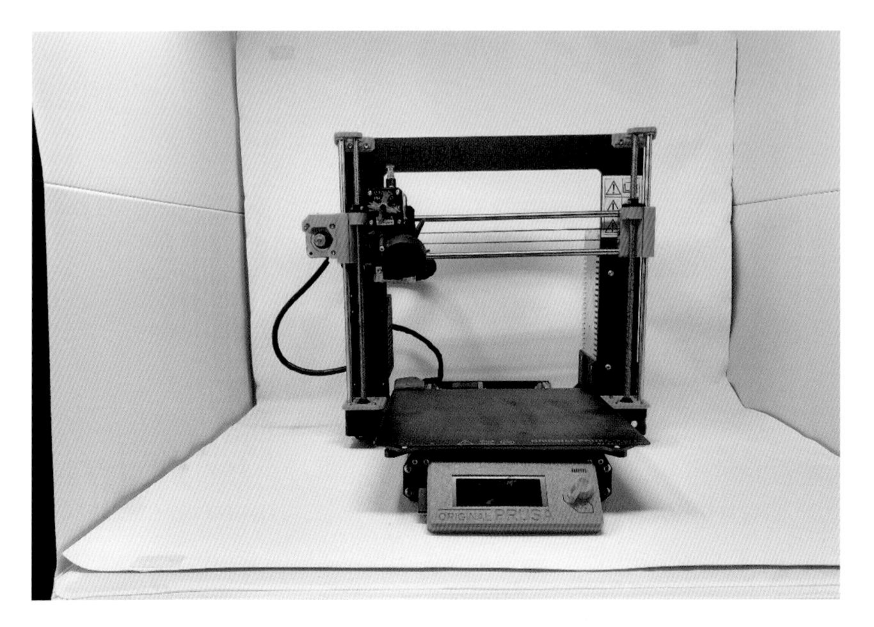

A photo box used as a printer enclosure

Extruder

The extruder pushes the filament into the hot end. It is made from a stepper motor and gearing, which drive the filament. It's a simple concept, but the correct implementation is essential to the operation of the printer. The controller runs the stepper to keep up with the printer for proper extrusion on the print head. When the end of a print line is reached, the extruder will run backward, pulling the filament back from the nozzle and creating retraction.

The hot end and the extruder are connected either directly or by using a Bowden tube. The Bowden tube is a Teflon tube with an inside diameter slightly bigger than 1.75 mm. The tube guides the filament from the extruder to the hot end, allowing the extruder to be mounted to the frame. There are advantages and disadvantages to both connection methods.

Just one example of an extruder

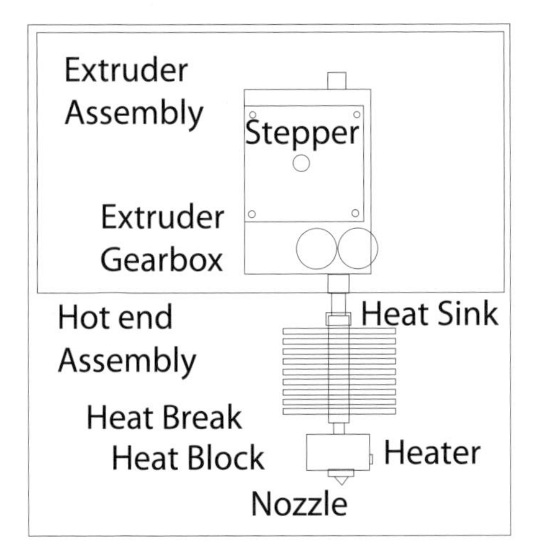

Basic extruder and hot end assembly

Having the extruder attached to the hot end increases the mass the printer needs to move around on the bed. This requires the whole printer to be stronger. Using a Bowden tube removes weight from the hot end but increases the distance the extruder must push the filament. The extruder needs to be able to compensate for the extra push and the flex of the filament inside the Bowden tube. Retraction settings are also higher when using a Bowden tube than when using a direct drive.

Hot End

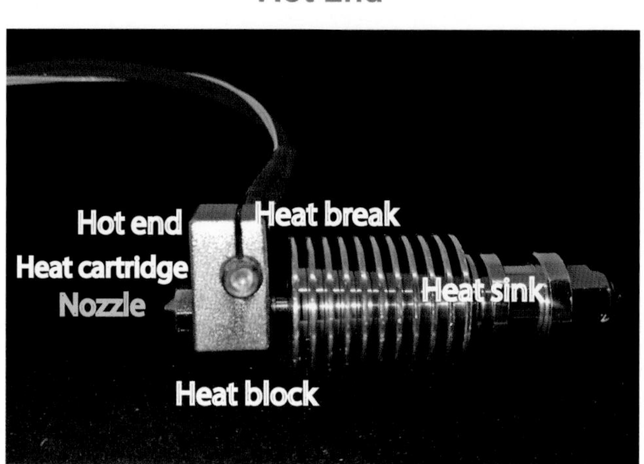

Hot end assembly

The hot end is the working part of the printer. The three main parts of the hot end are the heat block, heat break, and heat sink.

The heat block includes the nozzle, heater, and thermistor. The heat block is made out of aluminum or copper. Copper is limited to high-wattage hot ends, usually found on custom high-end printers.

Nozzles come in different sizes and materials depending on the material you are printing and the quality of the print. Most nozzles are brass, which works well for most filaments. The nozzle sizes start at 0.15 mm and go up to 1.2 mm. The 0.15 mm nozzle will work for very fine detail, but the print speed is slow. On the other hand, a 1.2 mm nozzle will print fast if you want a quick draft of a large piece.

Certain filaments, such as glow-in-the-dark or metal fill, are abrasive and will wear out a brass nozzle quickly. To prevent this, use a hardened-steel, stainless-steel, or ruby-tipped nozzle. The ruby tipped nozzles are very expensive and brittle, but they will last forever.

The heat cartridges come in 12 volt or 24 volt and will be 30, 40, or 60 watts. The higher wattage cartridges allow for faster heating of the heat block and faster heating of the filament, allowing for a faster print. Most printers use the 12-volt or 24-volt cartridges.

The heat sink includes a fan and keeps the filament cool until it hits the heat block. If the filament heats up too soon, it can jam up in the hot end, creating problems. Cooling becomes very important in the quality of the print. The smaller the area of molten filament, the better the quality of print.

The heat break is a piece that connects the heat block to the heat sink. It provides the separation between the molten and solid plastic.

Slicer

The slicer is the software program that tells the printer what and how to print a part. See chapter 6 to learn more about slicers.

Retraction

Retraction is the primary method used to start and stop printing. When printing, the hot end will lay down plastic until it gets to the end of a line. To stop the flow of plastic, the extruder is operated backward for a preset distance to stop the print. When the hot end gets to the start of the next line, the extruder operates forward to start the flow of plastic again. If retraction is working correctly, then the printer won't have stringing (threads of plastic between parts). For the most part, the settings for retraction from the factory will work fine.

Perimeters

Unless otherwise specified, all FDM printed parts are hollow. How hollow they are depends on the percent of infill and the number of perimeters. When the printer starts printing a layer, the first thing it does is to draw an outline of the layer. This outline is called the perimeter. Two perimeters are the usual, and minimum, number for most prints. To increase strength, you increase the number of perimeters and the infill percentage. For most prints two to four perimeters and 20% infill are a good place to start.

A single perimeter with no infill is called vase mode.

Skirt and Brim

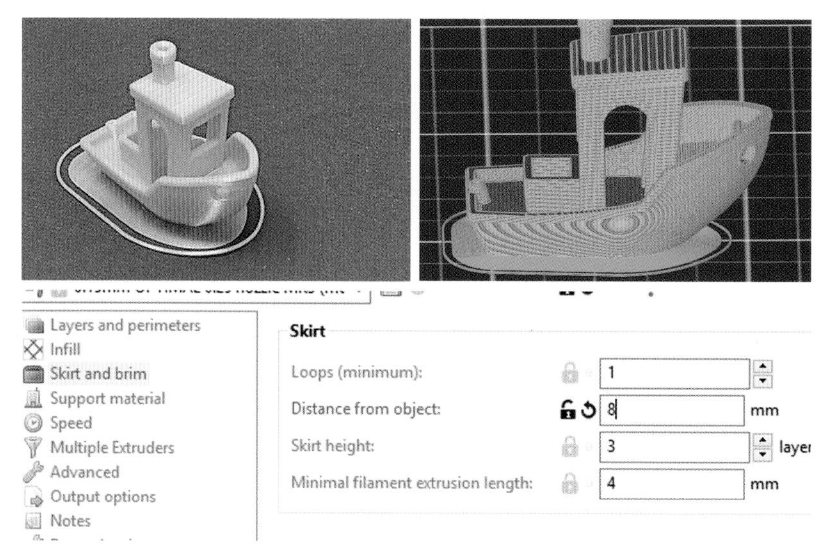

Benchy with skirt and brim

Skirts and brims are a process whereby the slicer prints an outline around the print. The skirt is outside the print, and the brim comes up to and connects with the base of the print. Skirts and brims provide a larger base for the part and create greater adhesion for tall, skinny parts.

Raft

A raft differs from a brim in that it is an independent structure, several layers thick. The raft is built under the print and around the outside. It provides a much greater level of bed adhesion for small prints.

Slic3r Prusa Edition - 1.41.2+win64
File Plater Object Window View Configuration Help
Plater Print Settings Filament Settings Printer Settings
0.15mm OPTIMAL 0.25 nozzle MK3 (mc
Layers and perimeters
Infill
Skirt and brim
Support material
Speed
Multiple Extruders
Advanced
Output options
Notes
Dependencies
Support material
Generate support material:
Auto generated supports:
Overhang threshold: 55 °
Enforce support for the first: 0 layers
Raft
Raft layers: 5 layers

Benchy printed with raft

Support

Supports are used when dealing with overhangs on the model. Printers can handle angles up to 45 degrees and some unsupported bridging, but models with cantilever designs need to be supported during printing. The slicing software will design removable supports when requested. In dual-material printers, the supports can be made from the same material as the print or by using soluble filaments.

Benchy with full supports

Infill

Infill is a support structure inside the print. Infill provides strength to a part while keeping the weight low and minimizing the cost of plastic. Available infill patterns depend on the slicer used. Most slicers provide a simple rectilinear pattern or a simple crosshatch. Others provide options that can provide additional strength in different directions using less plastic.

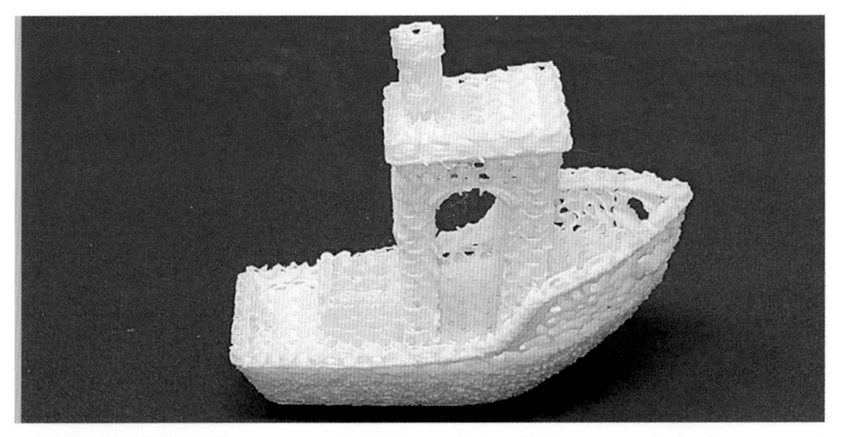

Benchy printed only with infill

A meat tenderizer printed using a dual-material printer and a gyroid infill, filled with a casting resin, and turned on a wood lathe

Chapter 5

What to Look for in a 3D Printer

Failure is always an option.

—Adam Savage, *MythBusters*

The items of interest in buying a 3D printer include build size, build plate, enclosure, hot end, extruder, and software.

Build size is the first consideration. How big are the items you plan to print? This is not always a question that is easy to answer. If your printing is going to be limited to small gears, drawer pulls, or bottle stoppers, then a printer with a 100 by 100 mm (4 × 4 inches) build plate will probably do everything you need. A larger print volume is nice, and build plates of 8 inches, 10 inches, and 12 inches square are not uncommon. Some inexpensive printers go up to 19 inches square. The projects designed in this book are built around a print bed of 220 by 220 mm (8.6 × 8.6 inches). All 3D printers are set up in millimeters. Projects need to be designed in or converted to millimeters or the print size will be wrong. So, make sure that your projects have been converted to millimeters before sending them to the slicer software.

The build plate needs to be perpendicular with the travel of the hot end. Depending on the printer, the bed may be self-leveling or manual leveling. Self-leveling is preferred over manual leveling. The important part is that the bed is flat, true, and level to the plane of the print head.

There are good 3D printers without a heated build plate, but they are limited to printing with PLA. If you are planning on printing using material other than PLA, a heated build plate is necessary to minimize warping and to create adhesion on the build plate. For ABS, the build plate needs to get up to 125°C to maintain adhesion and minimize warpage.

An enclosure is not needed, but helps control warpage and the cooling of material between layers. Drafts and breezes can cause temperature swings of the filament as it is being laid down on the print. This can cause warpage or failed layers. Many, but not all, printers come fully enclosed. For printers that do not come with an enclosure, there are frequently cases available from third-party vendors. These, or even homemade enclosures, will work as well as an enclosure built onto the printer. Heated enclosures are very rare in FDM printers because, unlike SLS printers, the plastics don't need to be maintained at temperature during the print. A simple enclosure can be made out of a couple of IKEA LACK tables and some plastic sheets, as shown in the picture below.

The hot end and the extruder are the components most critical for producing a good print. The problem is you can't tell a good hot end by looking at it. There are three ways to get an idea of the quality of the hot end and the extruder. The first is to ask the manufacturer for a sample print. Some will provide a sample, some won't. Samples represent the best the company can produce and will give you a

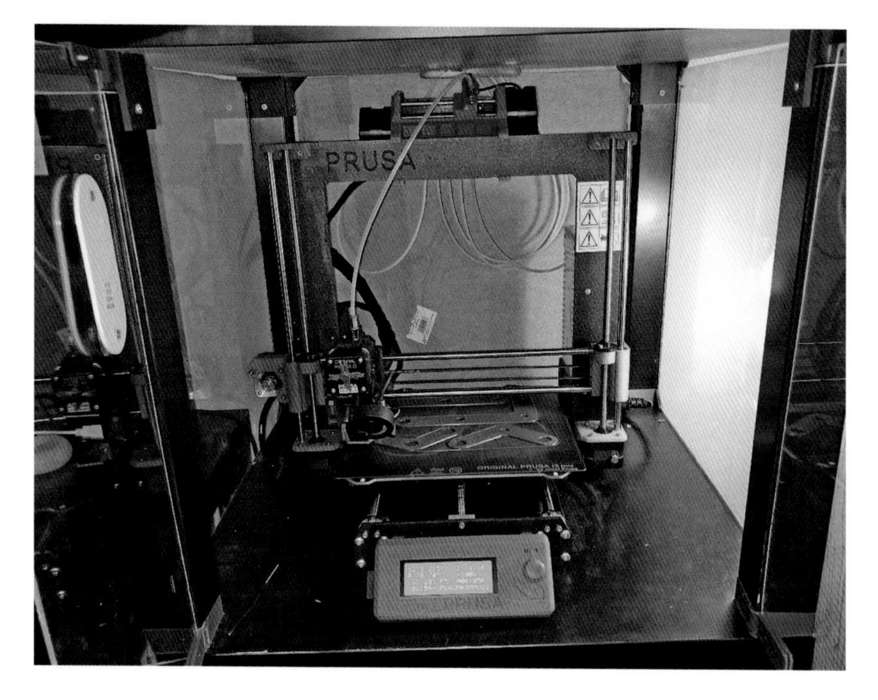

Prusa MK 3 printer with homemade enclosure

good idea of what is possible. The print they'll show you is probably post-processed, which means any trace of stringing will have been removed, but it still provides an indication of what is possible with a well-tuned printer.

The second way is to try to find a small print farm or maker space using the printers you are interested in. You can talk with the people there and order a small print without post-processing. The sample will cost you a nominal fee, but you will see what other users are getting out of their printers.

The third way is to look online. Several websites provide an annual review of printers and rate them by quality. ALL3DP.com, 3D Hubs.com, and *PC Magazine* also provide annual 3D printer reviews. You should take the reviews with a grain of salt, but they do have the latest information on the state of 3D printing. You can also check

The same file using two different printers

out YouTube for videos reviewing 3D printer equipment. There you can find a dozen or so 3D printing experts test out equipment that manufacturers have given them. Their reviews are honest and based on experience.

For the projects in this book you will need a printer with a 220 × 220 mm (8.6 × 8.6 inches) build plate that has a 220 mm (8.6 inches) build height or bigger. The printer also needs a heated bed capable of 100°C and a hot end capable of reaching 280°C.

One thing to watch out for is a printer that uses proprietary filament. Almost all PLA filaments use Natureworks 4043D resin as the base for the filament, so using a PLA that is different from that recommended by the manufacturer should not hurt the print head. Some printers limit you to a specific spool size or come with a warranty that limits you to using the manufacturer's own filaments.

This usually is done to increase the cost of the filament. Filaments can come in pellets, 1.75 mm and 2.85 mm in diameter. Most printers available to small businesses or hobbyists are 1.75 mm.

The three top-rated printers are the Cetus 3D printer for around $500, the Prusa original I3 MK for about $1,000, and the Ultimaker for over $10,000.

Tools

Here are a few tools you want around the 3D printer:

- a pair of tweezers, about 6 inches long, used to remove filament from the nozzle after a purge
- a pair of needle-nose pliers
- screwdrivers
- a metric Allen wrench set
- a metric star drive set
- a spatula for removing the print from the print bed
- a hot air gun
- a set of calipers for checking the dimensions on parts

Printer Setup

A 3D printer should be given its own space, such as in an enclosed cabinet or its own room. A printer is not like a toaster oven that can be tucked in a corner. You need access to all sides of the printer during normal operations.

All 3D printers make noise, some worse than others, so you are going to want to place the printer somewhere with some noise isolating qualities, preferably in a room that is not normally occupied but that allows easy access.

While you need to be physically at the printer to change the filament and remove the print, there is no need to be around the printer all the time. That's why internet control of the printer is a convenient feature. The print process can be monitored to a large extent via a webcam. This is especially helpful on long prints. With a webcam you can see if there are problems and stop the print remotely.

Not all printers have internet control. Webcam access is even rarer. OctoPrint is a free open-source option that runs on a Raspberry Pi computer. With a Raspberry, with a touch screen and camera, you can control your printer locally or remotely. And with a Pi camera you have a cheap way to watch the print.

Chapter 6

Software

That's the thing about people who think they hate computers. What they really hate is lousy programmers.

—Larry Niven

Nothing works today without software. Nowadays we have software or apps for most everything. 3D printing is no exception. Here's a quick overview.

The primary software used to drive the printer is the slicer software. The slicer does exactly what it says—it takes a 3D model and slices it into layers to be printed. Typically, the software lets you save multiple profiles based on the printer and filament. You can then choose attributes such as the quality of print, temperatures, infill, and other parameters. The software converts or compiles the STL file with the new parameters into the G-code used to drive the printer. The G-code includes information such as the bed temperature and the hot end temperature necessary for the filament being used. Other attributes include the number of top and bottom layers, the number of perimeters, and the type and density of the infill.

3D printers all come with their own specific firmware to run the printer. The firmware is set up with information necessary to read the G-code commands and operate the printer. This includes the print volume of the printer and any specific commands and controls needed to operate that particular printer.

Most slicer software is based on the Cura slicer engine, and is updated whenever the Cura engine software is updated (if the manufacturer decides this is needed). The manufacturer's software will provide all the settings needed to print. You will need to select only the material and the quality of the print. Most printers also come preprogrammed with an initial set of beginner projects which are designed to get the end user started as quickly as possible. Choose the project file and set it on the build plate. Pick the filament, usually from a drop-down list, and set the infill density. These simple-to-use programs do the job and allow you to get started making your own prints.

Other slicer programs are available. These programs provide a level of functionality beyond those provided by the manufacturer, and come with advanced features that utilize a greater degree of the machine's capabilities. The three most useful programs are Cura (mentioned above), PrusaSlicer, and Simplify3D. These programs provide full functionality in setting up and controlling the printer.

Using one of the open-source slicer programs, you can set up a profile for your printer that's different from the default one provided. This can be helpful if you want to create a profile with a different nozzle or different extrusion rates. A personal profile allows you to make a custom configuration if you have a custom-built printer. For example, you can generate custom filament settings if you are working with a new filament or just need to tweak some parameters in the current configuration.

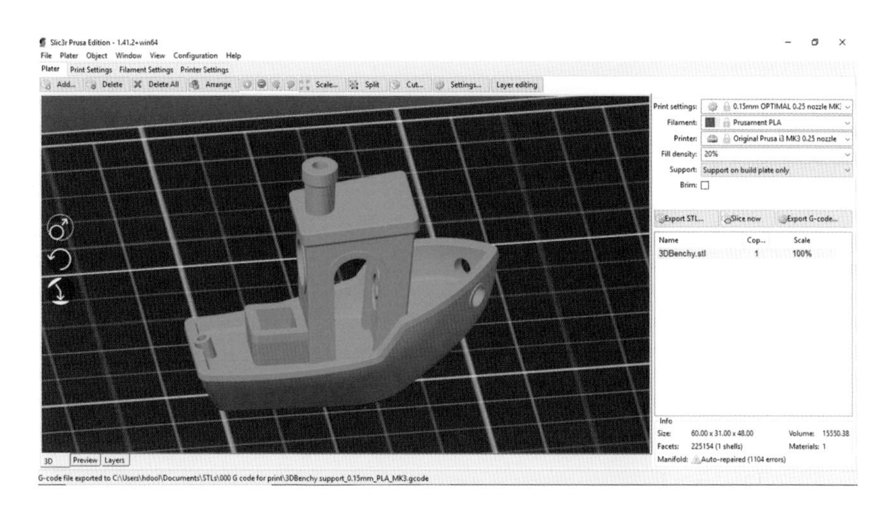

Benchy in PrusaSlicer software

PrusaSlicer is an open-source 3D printing program that gives you full access to all the parameters associated with your 3D printer. It was developed for the RepRap community and has drivers for most available printers. Don't confuse PrusaSlicer with Slicer. Slicer is a medical program that works with MRI images.

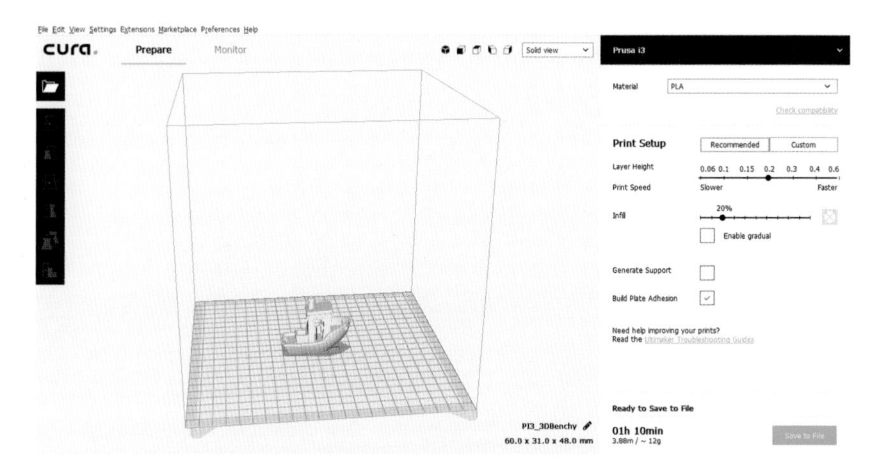

Benchy in Cura software

Cura, mentioned earlier, is a free 3D printing program developed and provided by Ultimaker. It is also open source and available for most printers. Cura is a highly capable slicer program that gives the user full control of the printer.

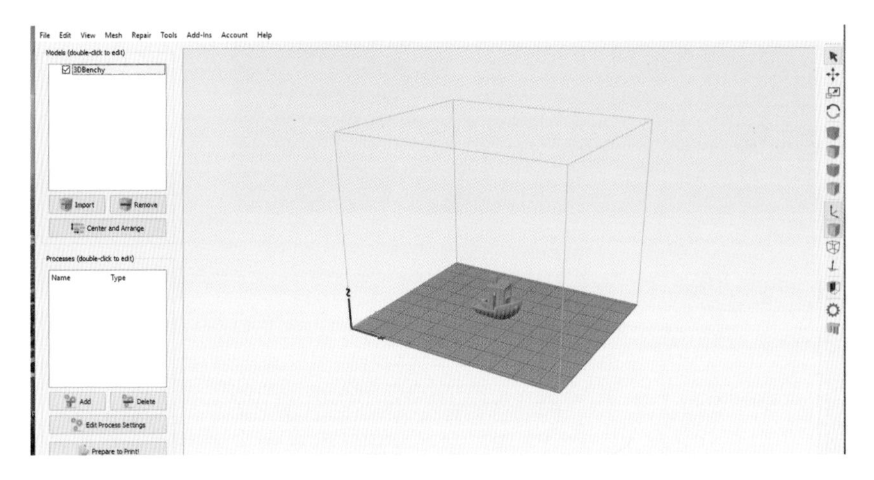

Benchy in Simplify3D software

Simplify3D is a commercial program for slicing with a few added features the other slicers don't have. With Simplify you can set different processes for different heights in the print. You may want low-density infill in the top of a print to save filament, but high-density infill in the bottom for strength. Simplify will allow you to do this. Simplify also has a repair function. If you bring a bad STL into Simplify, you can fix most problems without leaving the program.

Other open-source slicers are available that can be configured for most printers. Many of these slicers are based on Cura or PrusaSlicer. And they all have unique features. Some are designed primarily for a specific brand of printer with the idea that if you get used to their software, you will move toward the manufacturer's specific printer. Others, like MatterHackers MatterControl, are created with the idea of building customer loyalty toward their products.

Pathio is the latest slicer on the market. Like Simplify3D, it was developed from scratch as commercial software. Pathio's philosophy of slicing differs from that of its competition in that the outside edge is developed first to ensure that the part has a consistent thickness. Different slicers work better with different printers, and different people have better luck with different software.

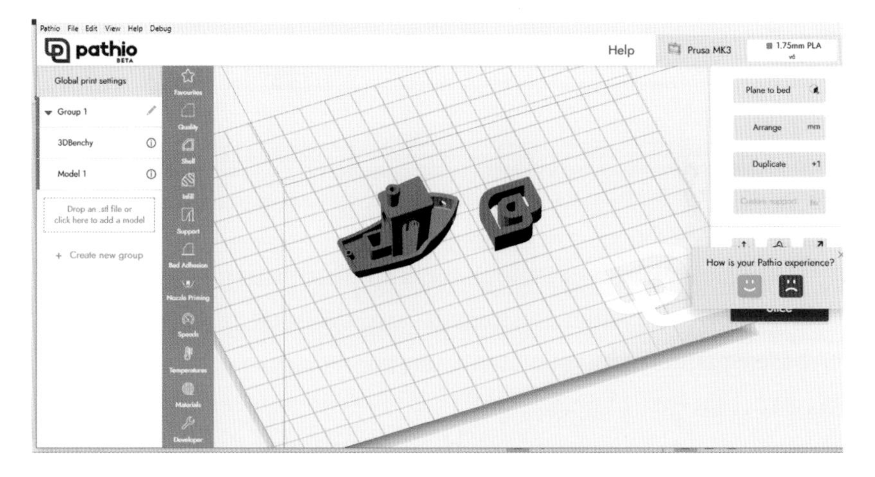

Benchy in Pathio software

Wi-Fi Connection

One of the newer additions to 3D printers is Wi-Fi. Some printers can be connected to the internet and controlled remotely. This has its advantages. You don't need to keep carrying a computer around with you to control the printer, or devote a dedicated computer for the printer. Printers do have a control panel associated with them, but these generally don't provide full control functions.

Having a remote connection to the printer with a camera attached allows you to remotely monitor the printer during a print and abandon the print should a problem arise.

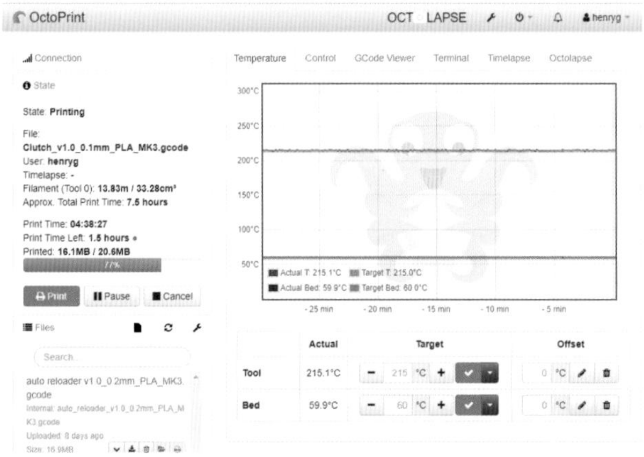

OctoPrint

Using a Raspberry Pi computer[2] and OctoPrint gives you remote control of your printer. OctoPrint is a 100% open-source, browser-based printer controller built for the sole purpose of providing Wi-Fi access for 3D printers. With OctoPrint you can use a webcam to monitor the printer and use hand controls on your computer or cell phone to control prints.

Different plug-ins allow you to customize your OctoPrint controller to your printer and your needs. Plug-ins are available for most printers. You can also collect statistics on your printer and print history.

As mentioned, OctoPrint runs on the Raspberry Pi computer. You can use any iteration of the Pi, but the Pi 3 has the most power. Some printers are set up to run on a Pi Zero W (a Wi-Fi enabled computer about the size of a stick of gum), so for about $9 you can set a printer up on the internet. Full instructions for downloading and setting up for OctoPrint are available at www.octoprint.org.

---

2. The Raspberry Pi is a $35 computer designed and manufactured for education. About the size of a deck of cards, it can be dedicated to the purpose of running a 3D printer, giving the user remote control and monitoring of the printer.

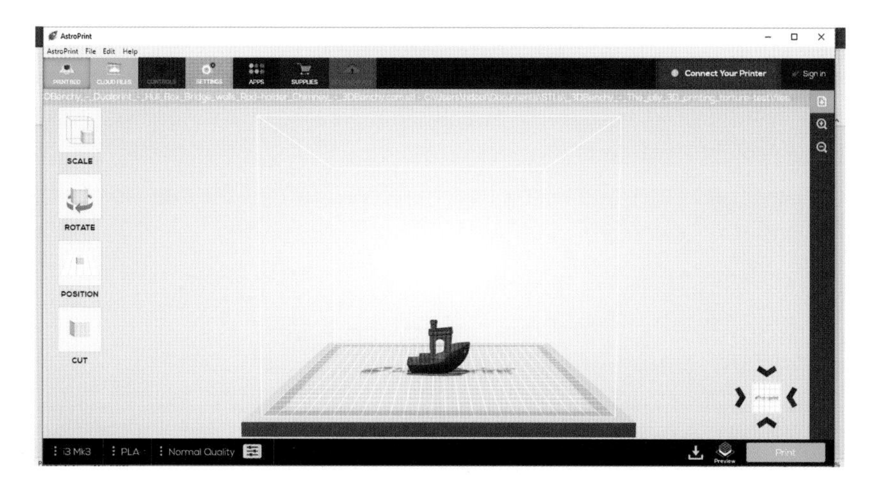

AstroPrint

OctoPrint is web based while AstroPrint is software based. AstroPrint can be set up and run from a home computer. Unlike OctoPrint, AstroPrint is made for profit. The basic control software is free for up to two printers, with plans available if additional printers are added. AstroPrint is designed to work with most printers. If your printer is not on the list, you can ask the manufacturer to add a profile. AstroPrint works as a rather simple slicer program with Wi-Fi capability. You can control your printer using your phone or tablet, along with your desktop computer, by adding a Raspberry Pi with AstroPrint software. AstroPrint is primarily aimed at Wi-Fi control of the printer and does not have the functionality of Cura or PrusaSlicer.

Toshiba FlashAir is another option for wireless operation of your printer. The Toshiba FlashAir is an SD card with Wi-Fi capability. By placing a FlashAir into the SD port on your 3D printer, you give it Wi-Fi capability. Once the card is installed you now need to program it for the printer and install software or configure the computer to run it. This is not that difficult, but the FlashAir costs about the same as building an OctoPrint system with a Raspberry Pi.

Design Software

Without projects to print your printer does you no good. Many sites have projects available that cover most needs. A lot of these project files are free, but some cost a nominal fee. Some of the sites to check are:

- www.thingiverse.com
- www.pinshape.com
- www.myminifactory.com
- www.youmagine.com

Not everything you are going to want to print is available online. The answer to this dilemma is to design your own projects.

Two basic philosophies or approaches are involved in CAD software. The first is solid objects. Designs are built using basic building blocks to achieve a final design. Boxes, cylinders, and spheres are joined together to build a part or parts. Holes are cut and filets are added for smoothing. The second method uses meshes. With meshes you sculpt the part. You can pull, push, add, or subtract edges and surfaces to create the desired form.

A number of different software programs are available for designing parts for 3D printing. They start with the very simple, like MatterHackers MatterControl, to the complex, such as SolidWorks—and all options in between. Many free programs are available, and all programs offer a free trial. The object is to find a program that is within your budget, is easy for you to understand, and will do what you need.

AutoCAD Inventor and SolidWorks are the two top-rated 3D CAD programs. Both programs have the ability to take a design and generate STL files for printing. Both have a high level of sophistication and are designed for use by professional draftsmen. A one-month free trial is available for each program, and a one-year subscription is available to veterans for the cost of shipping. If you are not a draftsman or are unable to hire one for your business, then SolidWorks or AutoCAD Inventor is probably not for you. Inventor, SolidWorks, and MatterControl use solid objects for design.

Meshmixer and Blender use mesh design. Both are free. Meshmixer is easier to use than Blender, which is rather high-end software.

AutoCAD Fusion 360 is unique in that it can use mesh or solid modeling to produce designs. The program is free for beginners, hobbyists, and small-business owners. Whereas training classes for SolidWorks are expensive, thousands of YouTube videos offer training for Fusion 360 for free. All the projects in this book were designed using Fusion 360.

Tinkercad is another free CAD program by AutoCAD. It's a good program for someone with no experience with CAD software. It provides the basic building blocks needed to design a project.

A lot of woodworkers are familiar with SketchUp. SketchUp is free and is another option for generating object or STL files for use with the printer.

The object is to find a program that you can live with and will do what you need.

Chapter 7

Filaments

There is a great future in plastics.

—Mr. McGuire, *The Graduate*

Filaments are plastics that have been formed into threads of specific diameters ready to be fed into the 3D printer. Filaments can be made from different plastics, and will have different characteristics depending on the plastic. Four things to consider when choosing a filament are the plastic glass temperature (or glass transition zone), the tensile strength of the material, material durability, and the ease of printing.

The plastic glass temperature refers to the temperature where the plastic starts to soften. It tells you how heat tolerant the plastic is. ABS has a high glass temperature and can be used in cars or other areas that see high temperatures. PLA has a low glass temperature, so parts made from PLA will not survive summer heat in a car.

Tensile strength indicates how much force the plastic can tolerate (or withstand). Will the part be placed under

Filament for 3D printers

a load? Nylons have a very high tensile strength. PLA, on the other hand, will deform under a load.

Durability or wear resistance is important for parts that are placed under constant use. A Lego will see very little wear use whereas a drawer or door handle can see quite a bit.

Ease of printing becomes one of the more important considerations for someone new to printing. PLA is one of the easiest filaments to print with. It has a low glass temperature and low tensile strength but doesn't have great durability. However, it is very easy to print with and is cheap.

In this chapter we will explore each of the most common types of filaments. The table below summarizes their main characteristics.

Table of Filament Properties (Strength, Durability, and Ease of Print Are Rated on a 5 Point Scale, with 5 Being the Best)

	ABS	PLA	PETG	Nylon	TPE	PC	Metal	Soluble
Print temperature (°C)	230-250	190-220	230-250	260-300	250-270	280-300	depends	
Strength	4	3	4	NA	2	5	4	1
Durability	4	4	4	NA	3	5	4	1
Glass temperature (°C)	105	60	88	70	50	110	depends	1
Ease of print	3	5	4	1	3	2	2	4

Acrylonitrile butadiene styrene (ABS)

Strength	4
Durability	4
Glass temperature	3
Ease of print	3

When you think of ABS plastic, think of Legos, which are made of ABS. ABS is the grandfather of 3D printing filaments. It is easy to print with and is a strong, durable plastic. The glass transition temperature of ABS is about 105°C, so it is ideal for high-temperature uses, such as the inside of a car. ABS is an amalgamation of three different plastics, so each manufacturer has a slightly different formula for making their plastic. Not all formulas will work equally well on all printers or applications. By changing ABS manufacturers, you may find that one works better than others for your application.

The problem with ABS is that it tends to shrink as it cools down. This can cause warpage of the prints. To prevent this, the printer requires a heated bed and an enclosed print area to print.

Polylactic acid (PLA)	
Strength	3
Durability	4
Glass temperature	2
Ease of print	5

Polylactic acid, or PLA plastic, is a thermal plastic made from corn. It is fully biodegradable to the point where it can be used as a temporary part in the human body, dissolving in about two years.

PLA is one of the most popular plastics in the 3D printing world. Not only is it biodegradable and made from a renewable resource, but its print temperature is low and it doesn't tend to warp like other plastics. The disadvantage of PLA is that it has low strength and tends to be brittle. PLA should not be used for projects that require impact resistance. PLA also has a low glass temperature, so it should not be used in environments where it will see high heat.

All manufacturers use just one resin to make PLA. Differences in filament color are due to the additives that manufacturers use.

Because of the lower print temperatures and the fact that it doesn't warp, PLA can be printed easily on low-end printers that don't have heated beds or enclosures.

Polyethylene terephthalate (PET or PETG)	
Strength	4
Durability	4
Glass temperature	4
Ease of print	4

PET is a plastic, and is the most common type of polyester used commercially. The G in PETG indicates that ethylene glycol is used in its production. All PET used to make filaments uses ethylene glycol and is PETG regardless of whether the G is present or not.

PET is also known as polyester, and 70% of the PET manufactured is used in the textile industry. Most of the remaining 30% is used in the manufacture of various containers and bottles. More PET is used in the production of rugs and leisure suits than in the manufacture of water bottles and straws.

PET is a little more difficult to print with than PLA, but plate adhesion is very good and warpage is not an issue. PET is stronger and more flexible than either ABS or PLA so it has less tendency to break. Most PET filaments will print between 220°C and 250°C, with some printing up to 270°C. Most PET filaments have a glass transition temperature of 80°C, making them better for use in hot environments than PLA, although not as good as ABS. Companies modify their PET filaments to achieve different results. PET is food-grade plastic. Some filaments have been developed for greater clarity, and there are PET filaments that can take higher temperatures.

Nylon

Strength	NA
Durability	NA
Glass temperature	5
Ease of print	1

Nylon is a generic designation for a family of synthetic thermoplastics developed in the 1930s by the DuPont company. Nylon is engineered plastic that is both tough and flexible. It was originally used for toothbrush bristles. Unlike other thermoplastics, many of the available nylons are based on chemical processes specific to the manufacture of that particular nylon, each one developed to exhibit different characteristics. Kevlar is a nylon thermoplastic.

The attributes listed above for nylon vary, and depend on the type of nylon you are using. Most nylons have high strength and durability, but some can be very flexible. Their glass temperature is high.

Nylon can be found with a spectrum of properties. Nylon filaments can be designed to print in a wide range of temperatures, so you need to verify a specific filament's print temperature before using it. Another factor is the varying amounts of carbon fibers in nylon filaments. The carbon can increase the strength and stiffness of the print. Nylon with carbon fiber can be as strong as aluminum.

A drawback to nylon is that it is difficult to get good adhesion on the build plate with it. Frequently, additives need to be used on the build plate, such as glue stick, to get the necessary adhesion. But one of the advantages of nylon is that it has good bonding between layers.

Thermoplastic elastomers (TPE)

Strength	2
Durability	3
Glass temperature	3
Ease of print	3

Thermoplastic elastomers are plastics that will retain an elastic behavior after being formed into a part. TPE is not practical for most printing, but there are applications where it is very useful, such as on the jaws of a clamp or vice. TPE prints much better in printers where the extruder is connected directly to the hot end, rather than those that use a Bowden tube system.

Polycarbonate (PC)

Strength	5
Durability	5
Glass temperature	4
Ease of print	2

Polycarbonate filaments are the strongest, toughest 3D printing filaments available. The glass transition temperature for polycarbonate is around 110°C. PC requires a very high print temperature, with most running around 280°C to 300°C. Some have been formulated to print between 250°C and 270°C, making them more compatible with current 3D printers. When you are looking for strength, PC and nylon are the two best options.

Metal-based filaments

Strength	4
Durability	4
Glass temperature	4
Ease of print	2

Like nylon, the attributes of the filament depend on the type of metal and the base plastic.

PLA filaments are available with added metal powders, including iron, copper, brass, and bronze. These filaments look like filler metal and can create some interesting effects, including the development of a patina on the project. A great use for metal-based filaments is creating faux metal fittings for embellishment of wood projects.

The problem with these filaments is that they are highly abrasive and can quickly erode a brass nozzle. Metal-based filaments should only be used only with a hardened steel nozzle.

Soluble/removable thermoplastics

Strength	1
Durability	1
Glass temperature	1
Ease of print	4

These filaments are intended for support only. The only attributes required of them are that they are easy to print with and the supports are easy to remove.

Anytime you have an overhang greater than 45 degrees you will need supports. Most slicer programs will design and install supports when needed, using the material in the extruder. With a dual-head or multi-material printer you have the option of using filaments that are specifically designed to work as support material. For example, PolySupport is a filament designed to be easily removed from the project, and is particularly useful for hard-to-access areas. Unlike if you use the same material for supports that you do for the rest of the print, PolySupport can be removed leaving little scarring on the print.

Soluble filaments can provide support in areas of the project that are completely inaccessible. You can print a bearing using soluble filament as a support material. The support material will separate the components, allowing the bearing to be printed in one part. The support material can be dissolved leaving a functional bearing.

In addition to the filaments discussed above, new filaments are being developed that are stronger, tougher, less prone to warpage, and easier to print. And cheaper and more user-friendly variations of the common filaments already in use are always being introduced. The most common and easiest to get started with is still PLA.

Filament Size

3D filaments come in two different sizes: 1.75 mm and 2.85 mm. The 1.75 filament is the most popular size. Make sure you have the correct size for your printer.

Chapter 8

The Print Process

The dilemma is how to use technology without losing touch with our craftsmanship. Or is that just cheating?

—Scott Grove, *Fine Woodworking*

With additive manufacturing, complexity is basically free, so any shape or grouping of shapes can be imagined and modeled for performance.

—Madhu Chinthavali, research scientist, Oak Ridge National Laboratory

We have the printer, so now what next? We need to find a place to set it up. The printer is going to be noisy. How noisy depends on the printer. And the printer is going to smell of plastic while it is printing. If you are printing filament made from the waste by-products of the beer or coffee industries (yes, I said beer or coffee), the smell can be rather nice. For all other filaments, the smell is not that nice. ABS is the worst. Recent studies have indicated that 3D printers put small amounts of materials into the air. This may not be too much of a problem but, as with all things, you should minimize your exposure.

You want to find a location for the printer that takes these issues into account. Try to place it in an area where you have a stable temperature and protection from drafts. It helps if your printer is in an enclosure. If you don't have an enclosure, you will need to maintain a constant temperature to prevent drafts from affecting the print. Having walls or barriers on three sides of the printer will help produce better prints.

Another consideration is you want an open work area, along with storage for your tools and filaments.

Most filaments are hydroscopic (they absorb moisture out of the air), particularly the nylons. You want to keep all open filaments in a dry, cool location. The amount of protection the filaments need depends on the humidity where you live. In Phoenix, Arizona, humidity isn't an issue so open filaments don't need a large amount of protection. An airtight Tupperware container is an ideal storage container. South Carolina and Florida have high heat and high humidity, mandating filaments be stored in containers with humidity control.

Several companies are selling heated storage/drying containers for filaments. In reality the same thing can be done using a food dehydrator and Tupperware.

For a new printer, there are calibrations that should be performed to ensure proper operation. These may include calibrating the extruder, the temperature, and the nozzle height above the bed. Perform all calibrations called for by your printer manufacturer.

Printers with manual leveling need to be leveled prior to the first print. Place the nozzle in one corner of the build plate and lower it to where it is just above the plate and you can slip a sheet of paper under it. Without changing the *z* height, repeat the steps for the remaining three corners.

For a printer with self-leveling capability, the process is much simpler. You may need to perform a mesh level in order to set the initial settings in the printer. Otherwise the printer will perform a self-level at the beginning of each print. Regardless, you need to follow your printer's directions for leveling the bed.

To install the filament into the printer, you first need to bring the hot end up to temperature. You want to start with PLA. PLA is cheap and easy to use. If you have a printer with the extruder attached to the hot end, then cut the end of the filament at an angle. This allows the filament to pass through the extruder and hot end more easily. If you have a printer with a Bowden tube, you want to cut the filament square. Cutting the filament at an angle risks cutting through the Bowden tube.

First Print

Set the first layer for your printer. Within the controls of your printer is a method for setting the first layer *z*. Set the first layer *z* and then run a print. Note that for all 3D printers, the unit of measurement is the millimeter. Slicing software does not recognize the English measurement system. If you make a cube one inch and import it into most slicer software, it will print out as a 1 mm cube.

Print the block, but stop after the first layer and check the quality of the print. Because the nozzle needs to push the filament onto the build plate, if the nozzle is too high the filament won't stick to the build plate. Rub your finger over the plastic and see if it stays attached to the build plate.

If the nozzle is too low, you won't get enough plastic onto the build plate. Either you will have no plastic or the plastic will smear out from under the nozzle. You want a nice clean first layer. In the

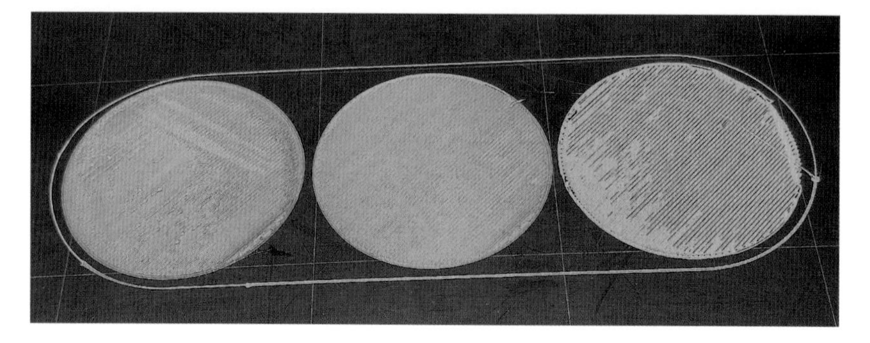

First layer differences in height

image below, the circle on the right is too high. The print on the left is too low. The one in the middle is the right height. If removed from the build plate, it will be smooth on both sides and stick well to the build plate.

The first three projects in this book are for basic shop tools. Performing these projects will test out the quality of your printer and give you an understanding of its features and capabilities. The basic calibration cubes should print in PLA with the basic settings used by the printer.

The first layer is printed at a lower speed and higher temperature in an attempt to get the necessary bed adhesion.

Before starting your first print, clean your build plate using soap and water and a Scotch-Brite abrasive pad. Completely dry the plate and then clean it off again with rubbing alcohol to remove any remaining oils and particles. After the initial cleaning, you can get by with cleaning the sheet every couple of prints with just rubbing alcohol.

This assumes that your printer uses a flexible PEI sheet. If your printer uses a nonremovable glass sheet, follow the manufacturer's directions for preparing the bed for printing.

1-inch cube

1-Inch Cube

First things first: you need to test your printer's calibration. Your first print after initial testing should be a 1-inch calibration cube. You can find 1-inch cubes on the internet or generate your own using any one of the free CAD programs available. Or you can use the file provided below.

Print the cube using the following settings:

File: www.thingiverse.com/thing:3535607
Material: PLA
Temperature: 200°C or in accordance with the manufacturer's directions
Bed temperature: 60°C (if you have a heated bed)
Infill: 20%
Support: none
Brim: not needed
Raft: not needed

After printing, use a micrometer to check the dimensions of the cube. If the dimensions are off, check with your printer manufacturer to see how to calibrate the printer.

1-inch die with 1 × 2 × 3 box

1 x 2 x 3 Box

The 1 x 2 x 3 project provides the same tests as the 1-inch cube but on a larger scale. Assuming the 1-inch cube printed correctly, you should not need to clean the build plate for a few prints.

As with the 1-inch cube, watch the first layer and adjust the *z* axis as needed to get a good quality first layer.

Print the 1 x 2 x 3 box using the following settings:

File: www.thingiverse.com/thing:3535607
Material: PLA
Temperature: 200°C or in accordance with the manufacturer's directions
Bed temperature: 60°C (if you have a heated bed)
Infill: 20%

Support: none
Brim: not needed
Raft: not needed

One of the concerns with 3D printing is under-extrusion. Under-extrusion is a condition whereby the printer is not pushing out as much plastic as the software expects. The software runs a line width calculated to join the extrusions, making a solid piece. If the printer is not providing the level of plastic required by the software, you get separation in the print, bad prints, and print failure. It would not hurt to stop a print before it's complete and look at the perimeters to see how well they are joined.

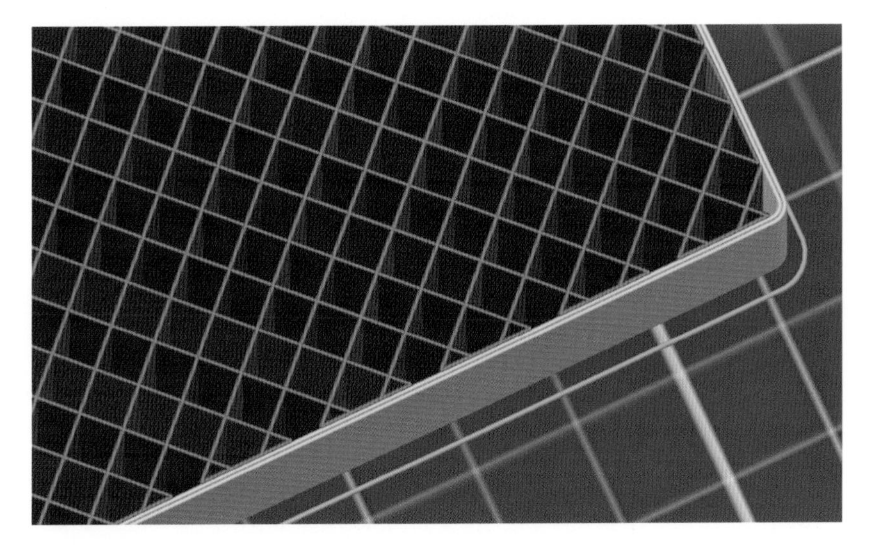

Infill with two perimeters

Torture Test

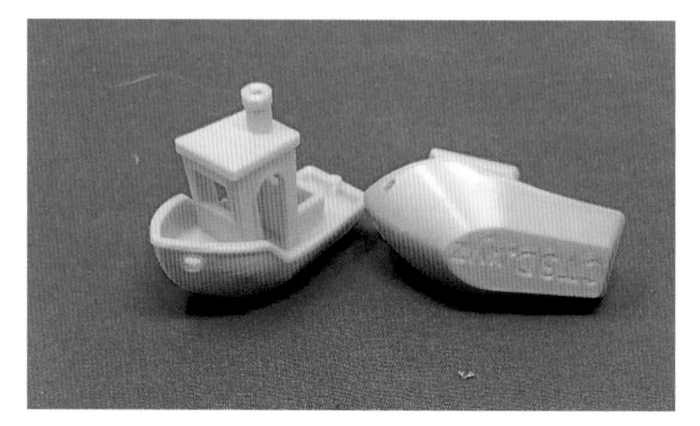

3D Benchy

Now that you have calibrated the first layer, it's time to check out the other capabilities of your printer. 3D Benchy is what is called a torture test. Developed by Creative Tools, it is one of the most commonly used torture tests available for 3D printers. The angles and dimensions on Benchy are all highly accurate and test the printer from different angles. With a good printer you can read the name plate on the back of Benchy.

Print Benchy using PLA and the standard settings in your printer and slicer. Different slicers will give different results, and we are looking to assess the quality using the settings that came with your printer. You can play with the settings later.

Print 3D Benchy using the following settings:

File: www.3dbenchy.com/download/
Material: PLA
Temperature: 200°C or in accordance with the manufacturer's directions
Bed temperature: 60°C (if you have a heated bed)
Infill: 20%

Support: none
Brim: not needed
Raft: not needed

Benchy stern

Benchy bow

When finished, compare your print with the dimensions at www.3Dbenchy.com/dimensions.

Benchy is designed to test your printer's ability to print circles and holes in different orientations and different sizes. This is helpful because a lot of tools you may want to print have holes, usually threaded.

Curves and angles and unsupported edges are important.

Benchy's bow runs out at the highest angle expected by a 3D printer. The windows test the printer's ability to print bridging and run filament across open spaces.

The dimensions for Benchy give you a very good indication of the precision of your printer. Specific dimensions can be checked to verify the accuracy of the printer, from the total length of the print to the diameter of the smokestack.

Benchy demonstrates the ability to print a first layer with letters. The best location to achieve high quality characters on a 3D print is on the bottom of the print up against the build plate. The plaque on the back of Benchy is 0.1 mm thick, offering the ability to show fine detail.

Chapter 9

What Can Go Wrong

1. If something can go wrong it will.
2. Murphy was an optimist.

Benchy, the good, the bad, and the ugly

As with all endeavors, there are obstacles to overcome. 3D printing is no exception. With some printers and some prints, everything will work perfectly. And it's nice to know that 3D printers are constantly being improved. The print heads are getting better and faster. Software is always being updated, and developers are finding ways to solve the problems and improve the prints. That said, there will always be problems to overcome.

The following sections of this chapter will help you identify the problems with your print and find a solution.

Layer Shifting

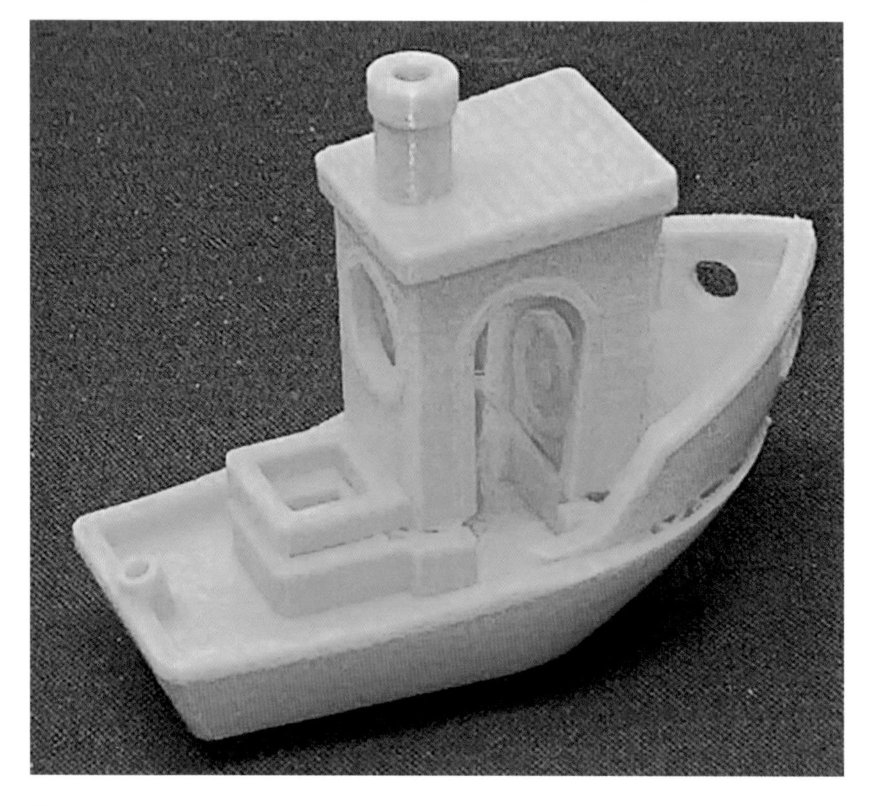

Benchy with layer shifting

Layer shifting is a problem whereby a belt shifts on a drive gear causing the print head to shift, or the build plate shifts on the heat bed. When this happens, your print is no good and you need to start over. There are several causes of this problem, but they are all mechanical. If layer shifting happens, look for obstructions that prevent the bed from full travel. Check the belts and belt drive gears. If these don't solve the problem, you may have a defective stepper motor.

Ghosting

Ghosting is a rippling effect on the surface of the print. For the most part, the effect is cosmetic in nature, but it can cause problems for tools used for marking or measuring. Ghosting is primarily caused by a lack of structural stability in the printer itself. The arm with the printer head moves too fast for the weight and starts vibrating. The problem can be mitigated by adding additional internal support to the print or slowing down the print speed.

Under-Extruded Prints and Skipped Layers

Under-extrusion and skipped layers are caused by either the extruder or the hot end. Any blockage in the nozzle can affect the flow of plastic out of the nozzle. Plastic will break down under heat over time. The length of time may not be all that long, but long enough to create a blockage in the hot end. The nozzle can be cleaned out using fine needles or using the cold pull method.

With the cold pull method, you heat the hot end and insert the filament. Turn off the hot end and let it cool down to between 140°C and 160°C, then disengage the extruder and pull the filament out. The filament should break off in the hot end and hopefully pull out the blockage. You may need to do this several times. Be sure to cut off the end of the filament each time or you may be pushing the blockage back into the hot end.

Under-extrusion can also be caused by problems with the extruder. The extruder is a mechanical device that needs to operate at a speed that matches the print speed. The gears in the extruder dig into the filament and drive it into the hot end. If the gears are dirty, they can't grab the filament. There are other problems with extruders that can cause under-extrusion. If your printer stops printing, the best course of action is to first clean the nozzle and then check the extruder.

The extrusion rate can be set in the slicer software. If the extrusion is set too low, you get under-extrusion; if it's set too high, you get over-extrusion. So confirm the extrusion rate is set correctly in the software before looking for mechanical issues with the extruder.

The difference between mechanical and software causes is that the software-caused issues are consistent through the print. The extrusion rate can be checked by taking 140 mm of filament, marking off 100 mm, and running it through the extruder. Remove the filament and verify that 100 mm had been used. If you have used more or less than 100 mm, adjust the extrusion rate.

Stringing

Stringing, or hairy prints, is when small strings of plastic are left when the extruder is moving to a new location during a nonprint move. Stringing is usually caused by improper retraction, overhangs, and bad filament. Improper retraction can be checked for within the slicer software, which has settings for retraction. But you want to be careful with these settings and make sure that retraction is truly the cause of the problem.

That said, it is difficult to prevent stringing with a Bowden tube printer. The flexibility of the Bowden tube, and the spacing between the filament and the Bowden tube, affect the extruder's control of the filament. Because of this the Bowden tube system requires longer movements of the filament to control stringing. If you have a Bowden

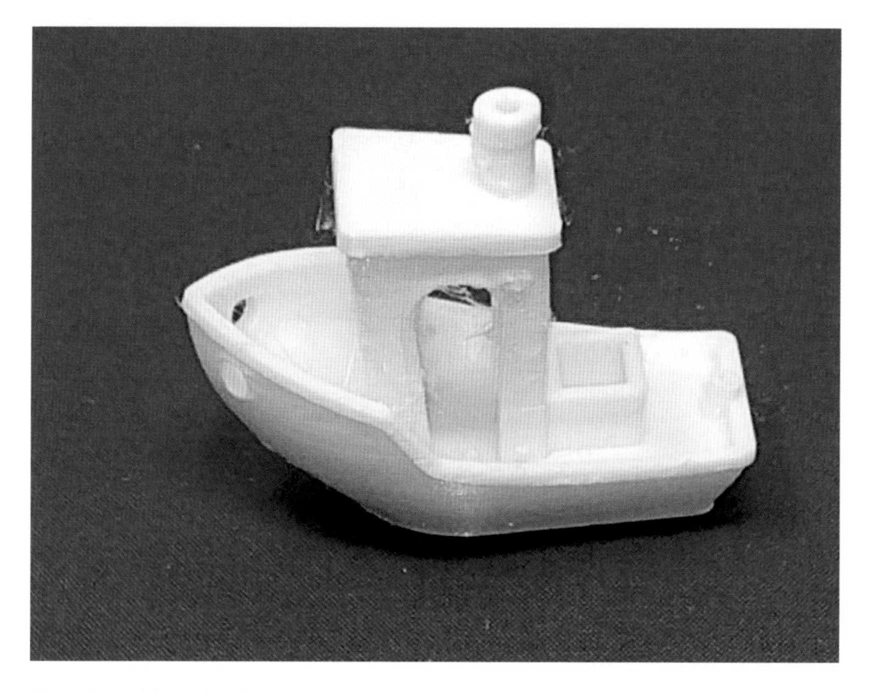

Benchy with stringing

tube printer and are experiencing a stringing problem, try a different slicer program. Cura's developers appear to have fixed the problem with stringing and Bowden tubes.

Overhangs and bad filament are also causes for stringing. Overhangs of less than 40 degrees tend to fail. This is a design issue that is easily corrected using supports. Supports are built into the slicer program and can be activated with the touch of a button.

There does not appear to be much bad filament sold in the marketplace. The cheap filaments work about as well as the more expensive ones. That doesn't mean you can't get bad filaments. The usual problems are diameter or moisture content. The industry standard deviation for 1.75 filament diameter is +/- 0.05 mm. If the filament is slightly out of spec, it can jam the printer, causing the printer to stop working. It is possible for a particular filament to work in one printer and not in another.

Old PLA filament can become brittle and break inside or outside the printer. If it breaks inside the filament path, the printer will stop and the filament will become difficult to remove. This filament likely has been sitting in a warehouse.

Just about all filaments are hydroscopic. They absorb moisture from the air. For this reason, they come in plastic bags with a desiccant to keep them dry. You need to keep the filament dry to ensure good prints.

Chapter 10

Print Projects

> 3D printing is already shaking our age-old notions of what can and can't be made. But what we've seen so far is just the tip of the iceberg. The next episode of 3D printing will involve printing entirely new kinds of materials. Eventually we will print complete products—circuits, motors, and batteries already included. At that point, all bets are off.
>
> —Hod Lipson, Director of Columbia University's Creative Machines Lab

You have a 3D printer. You have it put together and it is working. The printer has been tested and calibrated. Now what do you do with the printer? Here are some projects to help understand what the printer can do and how to get the most from your printer. The projects move from simple to complex, and will teach you important things such as how to orientate your project and when to use supports. They will also help develop your understanding of which filament will work best for each project and why. Let's go make something!

Paint Points

Paint points are used by woodworkers to hold up a project for painting, allowing for minimal contact with the part while applying paint. Due to its design, the paint points project introduces new challenges for your printer and unique elements for your prints. The paint points' walls are thin, their contact area with the bed is small, and the bottom of a paint point is curved. 3D printers don't do well with overhangs of less than 45 degrees. They will bridge between two points, depending on the distance between the points. The bottoms of the paint points are curved at an angle much less than 45 degrees, testing your printer's ability to print at a low angle. If the printer is having trouble with the angle and is leaving strings hanging below the curve, supports can be added. If supports are added in the slicer, the whole paint point will be supported, which is not the outcome you are looking for. The cure for this is to build supports into the design.

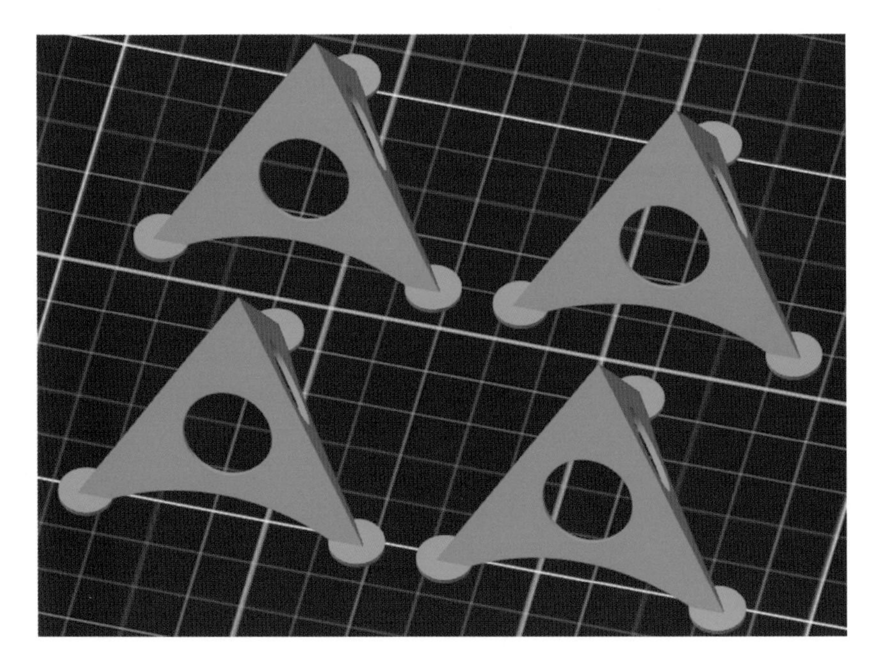

Paint points

Print the paint points using the following settings:

File: www.thingiverse.com/thing:3536003
Material: PLA
Temperature: 200°C or in accordance with the manufacturer's directions
Bed temperature: 60°C (if you have a heated bed)
Infill: 20%
Support: none
Brim: not needed
Raft: not needed

Paint points ready for use

Center Finders

Center Finders for Dowels

Center finders are used to quickly find the center of a dowel or block of wood by tracing a 45-degree angle across the wood. When printing the center finders for dowels project below, we need to be concerned about the placement of the part on the build plate. There are three flat sides to this project, but two of them have unsupported edges. If either of these two sides is placed on the build plate, without supports, the print will probably fail. Supports use material and cause waste, so we prefer to avoid using them unless necessary. We don't want to use supports with this project, so we set the orientation on a plane that doesn't have unsupported sides.

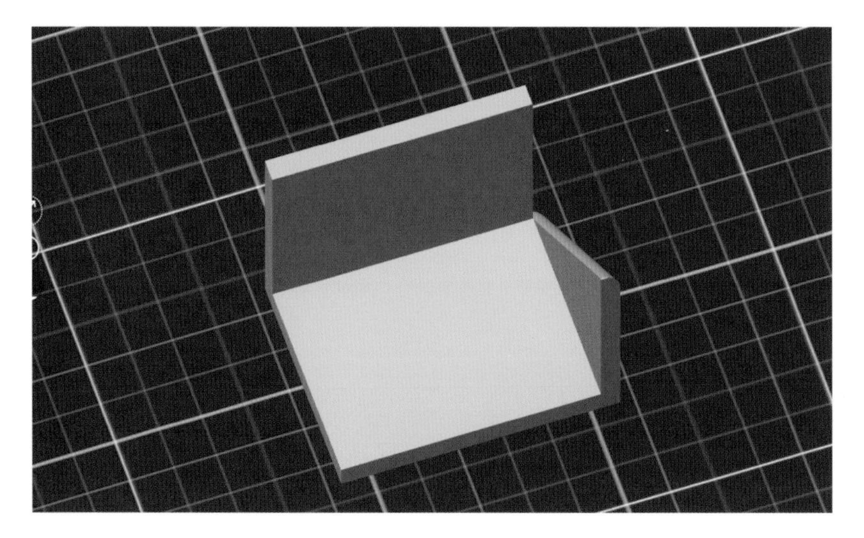

Center finder for dowels

This center finder is primarily designed for use with dowels and 2x square stock, but can be used for larger pieces.

Print the dowel center finder using the following settings:

File: www.thingiverse.com/thing:3536035
Material: PLA
Temperature: 200°C or in accordance with the manufacturer's directions for PLA
Bed temperature: 60°C (if you have a heated bed) for PLA
Infill: 20%
Support: none
Brim: not needed
Raft: not needed

Center Finders for Bowl Stock

This is a larger center finder that is primarily designed for use with round or square stock. The center is angled at 45 degrees. Most printers will print unsupported at this angle without sagging. The project can be printed in any one of the three orientations. The strength of a print will vary in different directions depending on the orientation of the layers. Think of the printed plastic as having a grain structure not unlike that of wood. On the whole, this is an important consideration in 3D printing, although it is not important in this project. One of the nice things about this design is that it can be modified with a company logo.

Print the center finder using the following settings:

File: www.thingiverse.com/thing:3536035
Material: PLA
Temperature: 200°C or in accordance with the manufacturer's directions for PLA
Bed temperature: 60°C (if you have a heated bed) for PLA
Infill: 20%
Support: none
Brim: not needed
Raft: not needed

Center finder for bowl stock

Center finders

Measuring Instruments

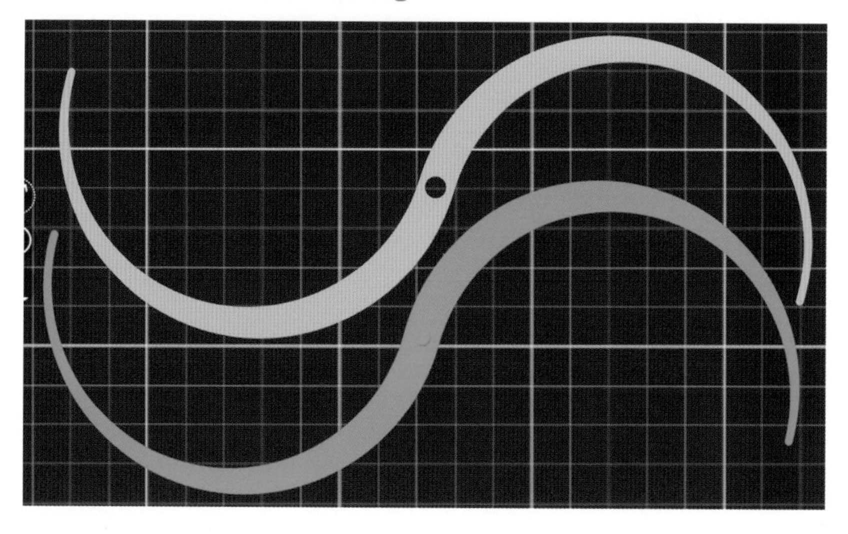

Bowl thickness calipers

3D-printed calipers

Print the calipers using the following settings:

File: www.thingiverse.com/thing:3536043
Material: PLA
Temperature: 200°C or in accordance with the manufacturer's directions for PLA
Bed temperature: 60°C (if you have a heated bed) for PLA
Infill: 20%
Support: none
Brim: not needed
Raft: not needed

Speed Square

Print Adhesion

I tend to keep two or three speed squares around the shop and am constantly looking for them. You can get a good speed square for $5, but with a 3D printer you can print one anytime you need it for $1. The speed square should be printed upright. Printing upright allows the speed square to have a full base, giving you a 45-degree angle in either direction. To accomplish this, you need good adhesion. The height of the speed square combined with the small base will give you a good indication of the adhesion you have on your build plate.

I find I get a little more strength if I use four perimeters and 40% infill. Not all slicer software will allow you to change the number of perimeters.

Printed speed square

Print the speed square using the following settings:

File: www.thingiverse.com/thing:3536050
Material: PLA
Temperature: 200 C or in accordance with the manufacturer's directions for PLA
Bed temperature: 60 C (if you have a heated bed) for PLA
Infill: 40%
Perimeters: 4
Support: none
Brim: not needed
Raft: not needed

Marking Gauge

First Threads

The marking gauge project includes our first threaded part. Threaded parts are a unique challenge in that the part needs to be printed in the correct orientation for the threads to work.

The marking gauge comes with a threaded screw to hold the slide in place. The screw is a standard ⅜ thread. Some printers are able to print the screw and thread hole without any adjustment. If the screw doesn't fit the first time, a 2% size reduction should work. For inside threads, the part can be printed in most any orientation and a good 3D printer will print acceptable threads. If the body is printed upright, it will need supports on the inside. For outside threads, as on the bolt, the threaded section needs to be printed standing upright. You can't print threads horizontally. It doesn't work. By printing the bolt vertically, the printer generates the threads moving up the bolt without supports.

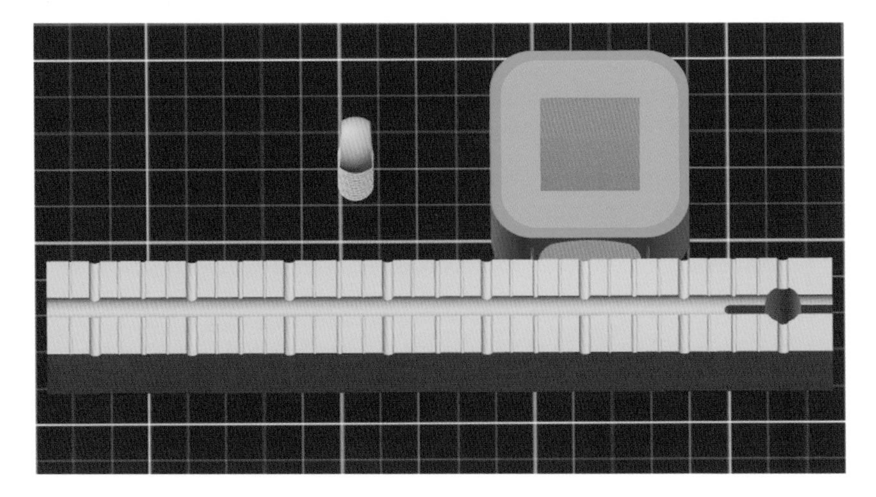

Marking gauge print layout

The tolerance between the body and the slide is 0.25 mm. Most good printers can get down to a tolerance of 0.15 mm or less. If body and slide don't fit, you can bump up the size of the body to get it to work.

Print the marking gauge with the following settings:

File: www.thingiverse.com/thing:3536065
Material: PLA
Temperature: 200°C or in accordance with the manufacturer's directions for PLA
Bed temperature: 60°C (if you have a heated bed) for PLA
Infill: 20%
Support: none
Brim: use with nylon to provide stability
Raft: not needed

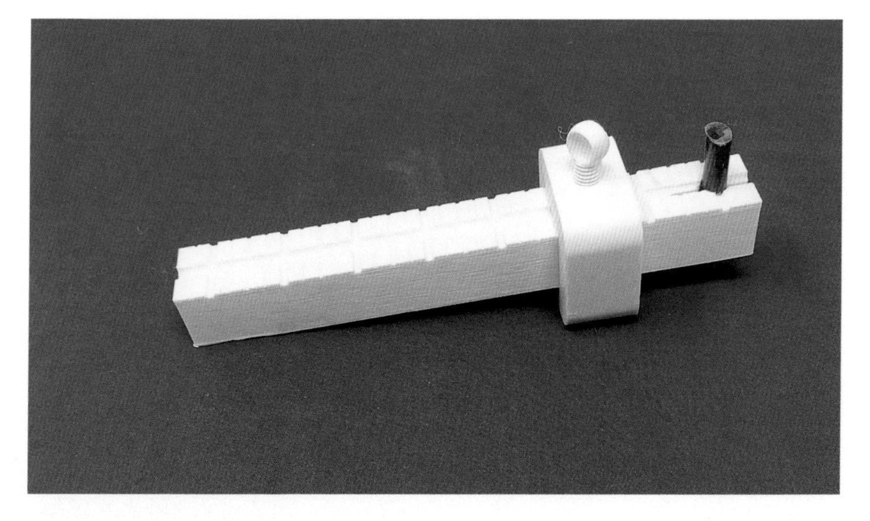

Printed marking gauge

Pocket Screw Jig

Current pocket screw jigs have a metal insert to prevent wear. PLA will work if you are not making too many holes. Otherwise there are other thermoplastics that have a much higher wear resistance. Polycarbonate, nylon, and PET/PETG all have a higher wear resistance than PLA and are better choices for this print. PET adheres to a PEI sheet better than PC and much better than nylon.

Nylon is very wear resistant and slightly flexible, allowing some give during use. As long as you have good layer adhesion the nylon will not break. It will bend and stretch under stress, but it will hold.

PC is stiff and hard. It is also very strong and wear resistant. PC won't last as long as a Kreg pocket hole jig with a metal liner, but it should outlast most jobs.

For your first print of this project, I recommend going with PET. It's stronger and more wear resistant than PLA and almost as cheap.

This pocket jig is set up for a specific angle and drill size, but using a 3D printer you can make jigs for any drill bit at any angle.

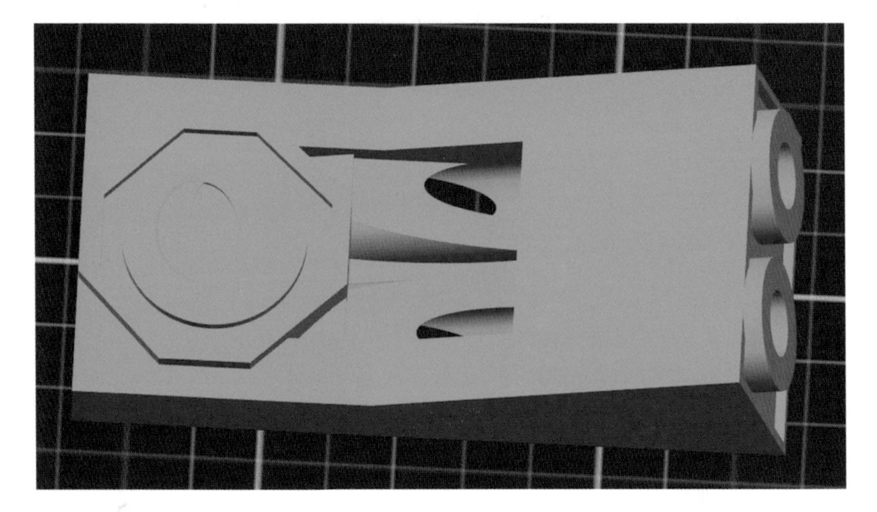

Basic pocket jig

Print the pocket jig using the following settings:

File: www.thingiverse.com/thing:3536109
Material: PETG
Temperature: 230°C or in accordance with the manufacturer's directions for PETG
Bed temperature: 85°C to 90°C for PETG
Infill: 20%
Support: none
Brim: not needed
Raft: not needed

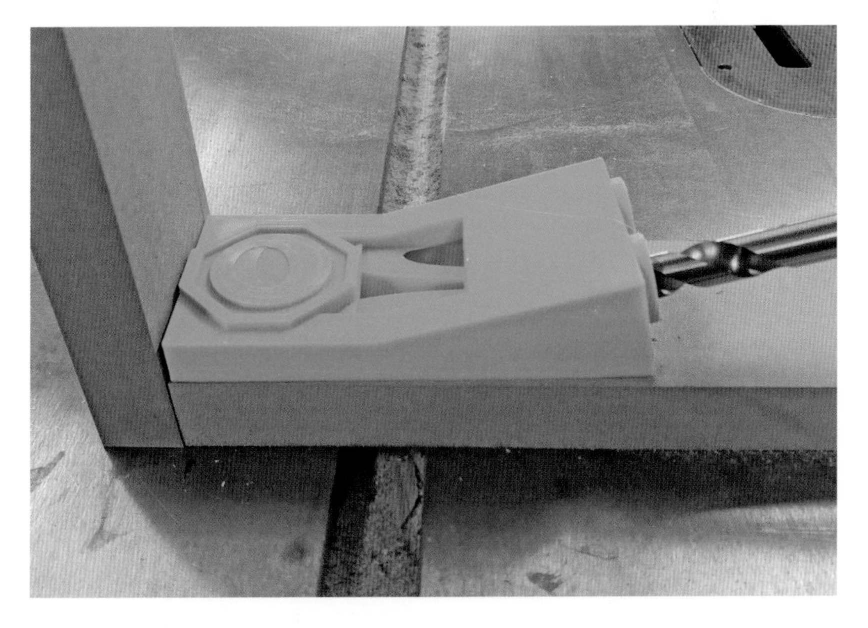

Printed pocket screw jig

Center Drill Jig for Dowels

Rafts and Supports

The design of this print has very little contact with the build plate, and the angle at the top of the print will have problems if it is not supported. A raft and supports should be used with the print. The raft spreads the contact area out and provides a removable base for the print, ensuring adhesion with the build plate. The disadvantage of the raft is that it uses more filament and creates some distress with the contact area on the print.

Supports allow you to print overhangs that cannot be printed without help. The slicer software will generate the raft and supports when told to. You only need to decide when the supports are necessary and how to remove them.

This part could be printed on its side with no supports, but what is the fun in that?

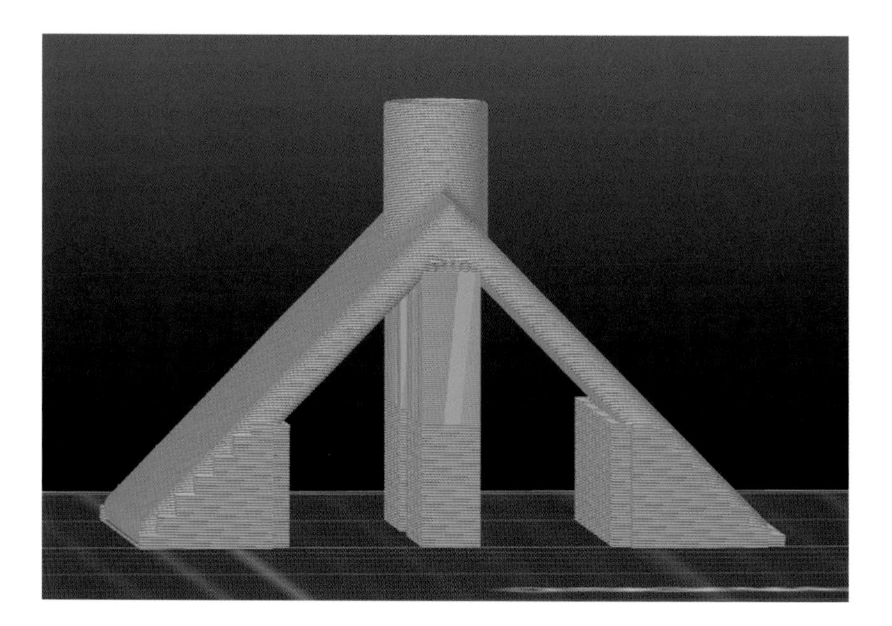

Center drill jig

Print the center hole jig using the following settings:

File: www.thingiverse.com/thing:3536093
Material: Nylon
Temperature: 255°C or in accordance with manufacture directions for Nylon
Bed temperature: 85°C to 90°C
Infill: 20%
Support: set supports
Brim: not needed
Raft: not needed

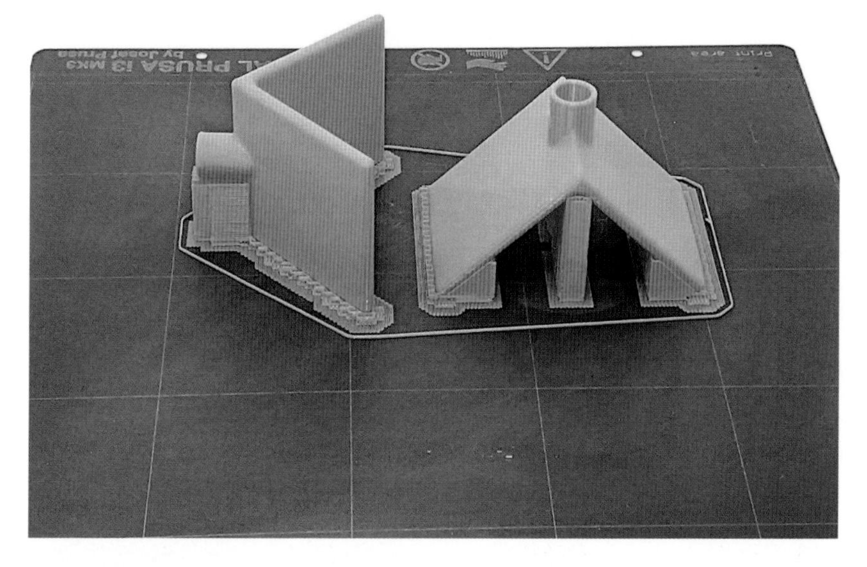

Printed center drill jig with supports

Router Jigs

Router bits designed for use with templates can cut a variety of shapes so long as the correct template is available. Using a 3D printer, you can make templates for any design you might want. Corner templates are popular. Pin routing jigs can be made on a 3D printer, giving you unlimited potential.

Here are three round corner templates with ½-inch, 3/8-inch, and ¼-inch radiuses. I consider the router templates to be a single project. The templates can be made for mortise-size jigs, pin-center mortising jigs, and zero-clearance throat plates for router bases, table saws, and band saws.

Print the template using the following settings:

File: www.thingiverse.com/thing:3536123
Material: PLA
Temperature: 240°C or in accordance with manufacture directions
Bed temperature: 85°C for PETG
Infill: 20%
Support: set supports
Brim: not needed
Raft: not needed

Printed mortise jigs

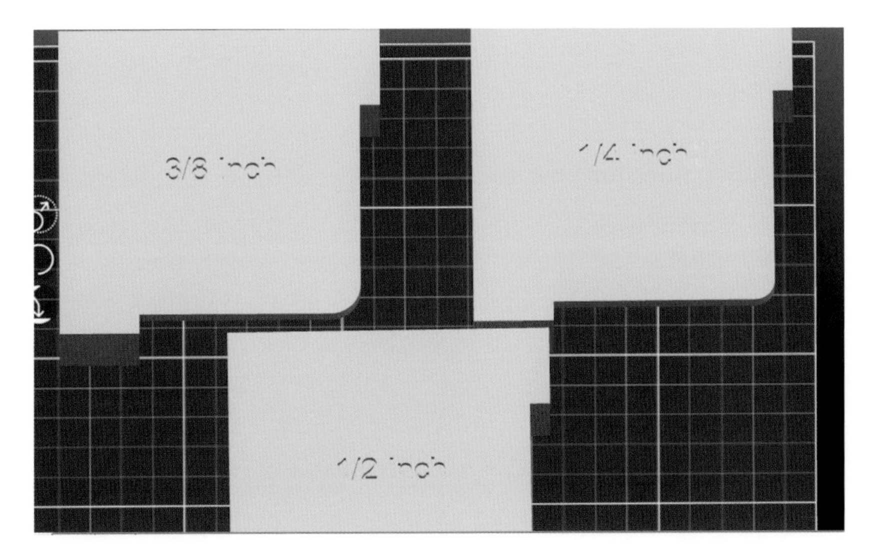

Mortise-size jigs, pin-center mortising jigs, and zero-clearance throat plates for router bases

Boxes

Boxes, jars, and containers are in constant demand in any shop. 3D printers can print a jar of any size within the range of the printer. Jars can be made to hold nuts, bolts, nails, or other pieces of hardware. Boxes can be made to hold custom projects. Adding threaded lids or hinges to boxes is easy using the printer.

Two projects are provided, one a round box (a jar), the other a hinged box. The hinged box project demonstrates the capability of additive manufacturing to create parts within parts. The box hinge pin is printed within the hinge and at the same time as the hinge. The hinge pin is free floating but not removable from the hinge. The ability to print parts within other parts is unique to 3D printing. No other technology available to the small shop has this capacity.

Print the box using the following settings:

File: www.thingiverse.com/thing:3536137 (simple lidded jar)
www.thingiverse.com/thing:3536149 (hinged box)
Material: PLA
Temperature: 215°C or in accordance with the manufacturer's directions for PLA
Bed temperature: 60°C for PLA
Infill: 20%
Support: set supports for hinged box; supports not needed for the round box
Brim: not needed
Raft: not needed

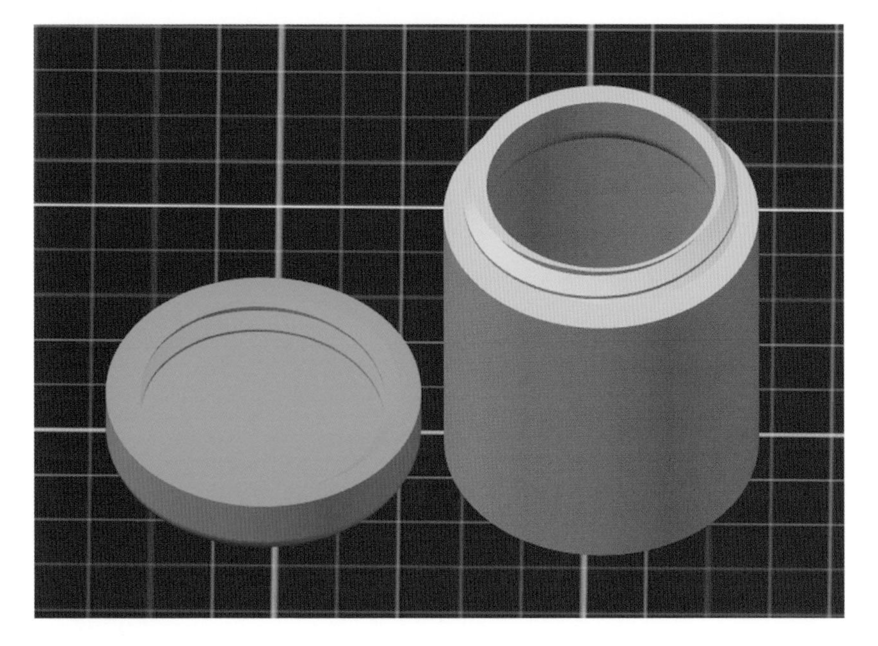

Pill box (jar)

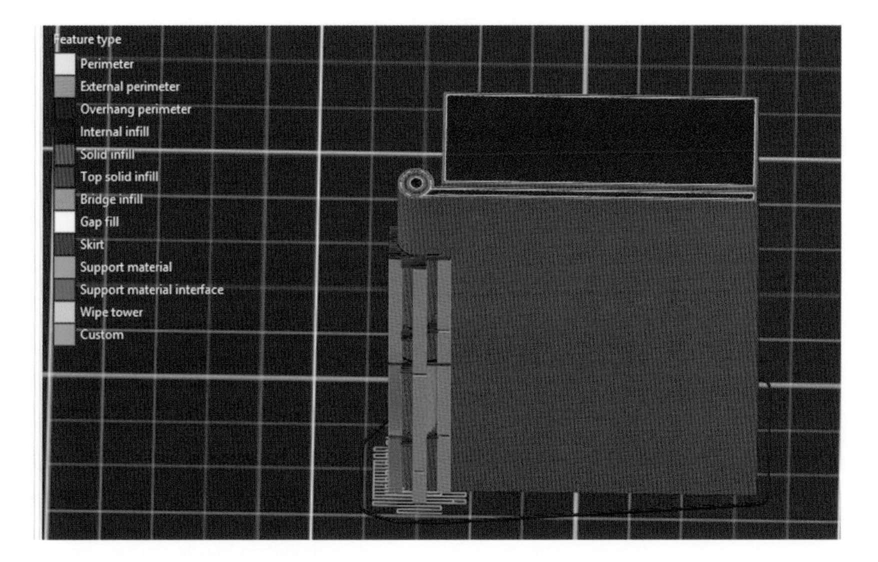

Hinged business card box

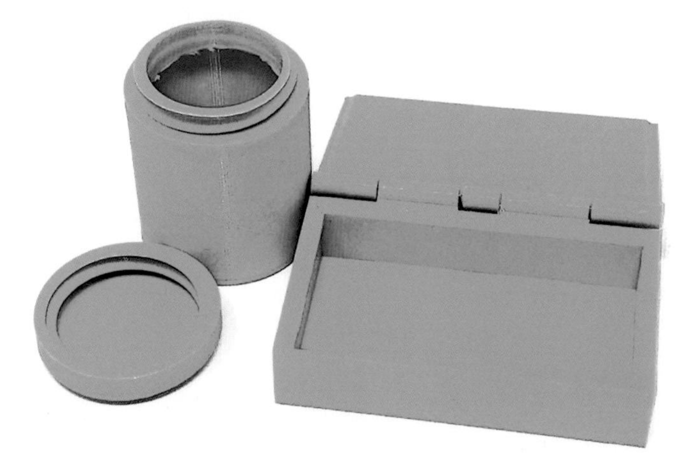

Printed boxes

Corner Jig

The strength of a part is primarily determined by the filament, wall thickness, and infill. Although most slicer programs allow you to change wall thickness, not all do, leaving you with just filament and infill. Strength can also be changed in the design process.

A corner jig needs to be strong. This design adds extra holes and walls to increase the strength and reduce the flexibility of the jig. The design works the same as cross bracing in a house to increase the strength of a wall. When you add to a design, you increase the number of walls or surface area. The slicer will then add perimeters around each hole, thereby strengthening the part. The holes can be extremely small and still do the job of adding strength to the print. Adding holes to a part does not necessarily reduce the amount of filament used.

Print the corner jig using the following settings:

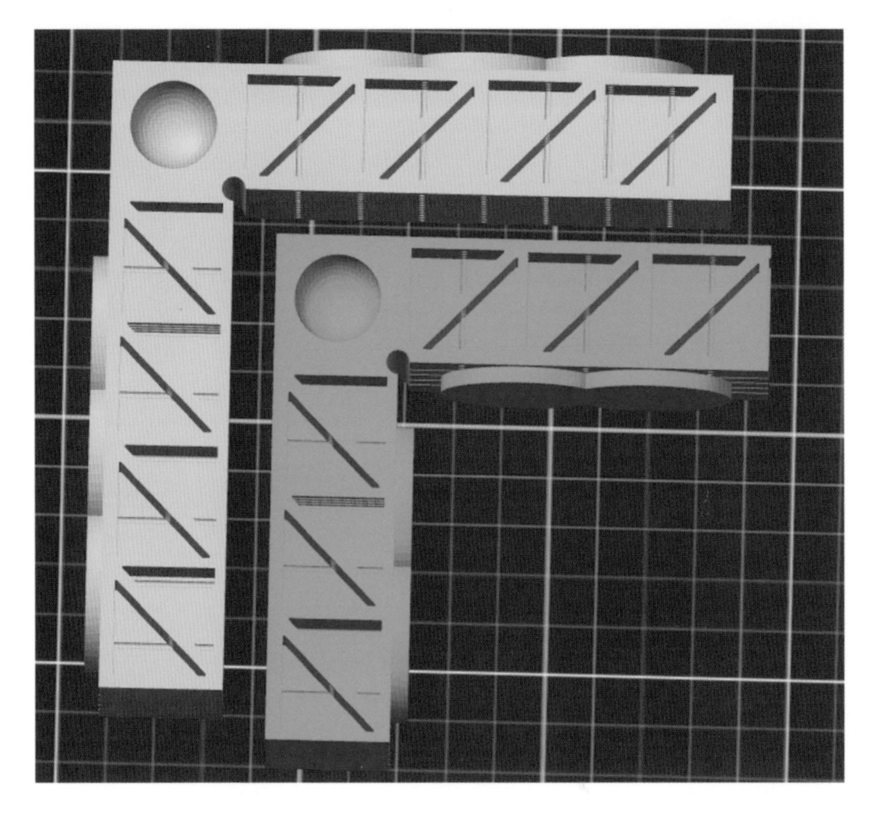

Corner jigs for clamping

File: www.thingiverse.com/thing:3536160

Material: nylon

Temperature: 255°C or in accordance with the manufacturer's directions for nylon

Bed temperature: 85°C to 90°C for nylon

Infill: 20%

Support: set supports

Brim: not needed

Raft: not needed

Printed corner jig

Threaded Chuck Holders

Anyone who uses a lathe will have multiple chucks and face plates. A simple threaded holder can help to organize the chucks and keep them handy.

Print the threaded chuck holder using the following settings:

File: www.thingiverse.com/thing:2422549 (English thread, 1.25 x 8)
www.thingiverse.com/thing:2422546 (Metric thread, M33 x 3.5)
Material: nylon
Temperature: 255°C or in accordance with the manufacturer's directions for nylon
Bed temperature: 85°C to 90°C for nylon
Infill: 20%
Support: set supports
Brim: not needed
Raft: not needed

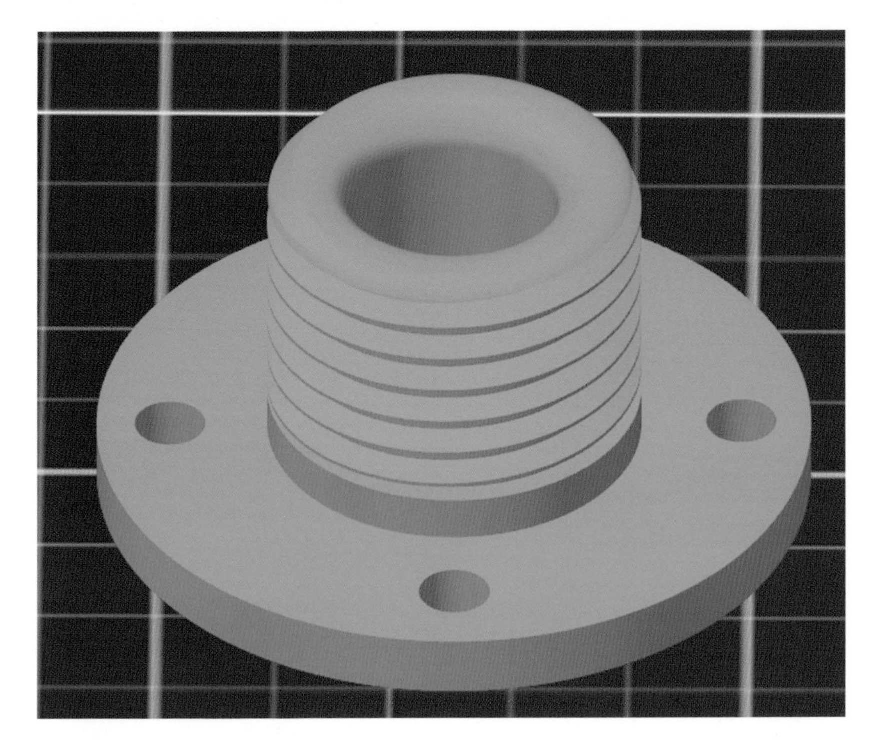

Holder for spare chucks

Printed chuck holders

Dust Collector Hood

Print the dust collector hood using the following settings:

File: www.thingiverse.com/thing:3536170
Material: nylon
Temperature: 255 C or in accordance with the manufacturer's directions for nylon.
Bed temperature: 85°C to 90°C for nylon
Infill: 20%
Support: set supports
Brim: not needed
Raft: not needed

The print will take about 8.5 meters of filament (25.5 grams), cost about 65 cents, and have a print time of about one hour. Time and cost are dependent on the filament and printer used.

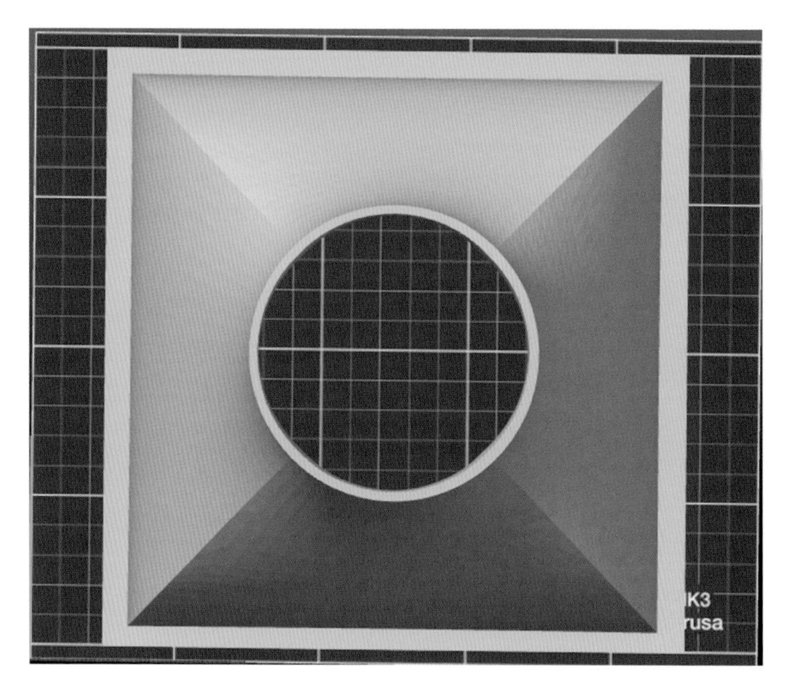

Dust collector hood

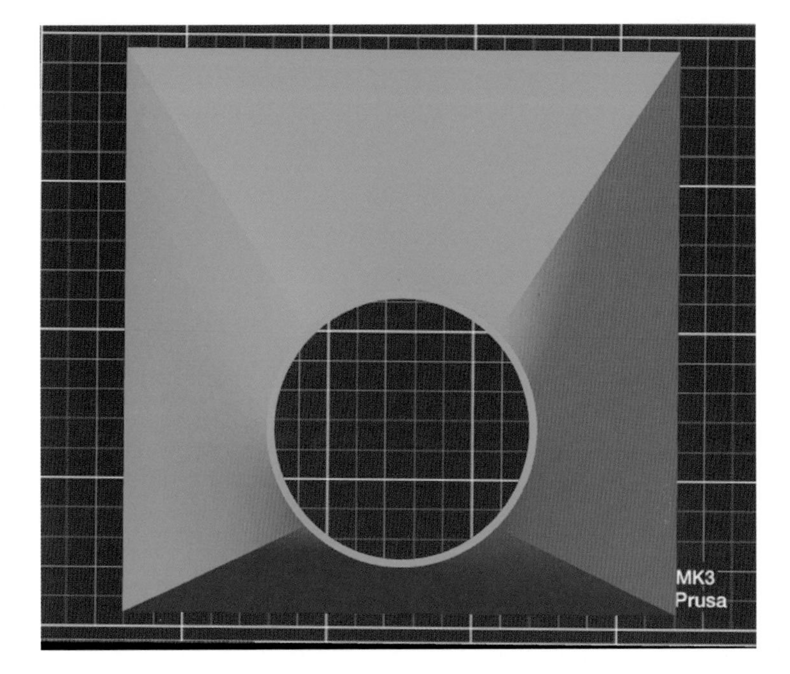

Dust collector hood offset

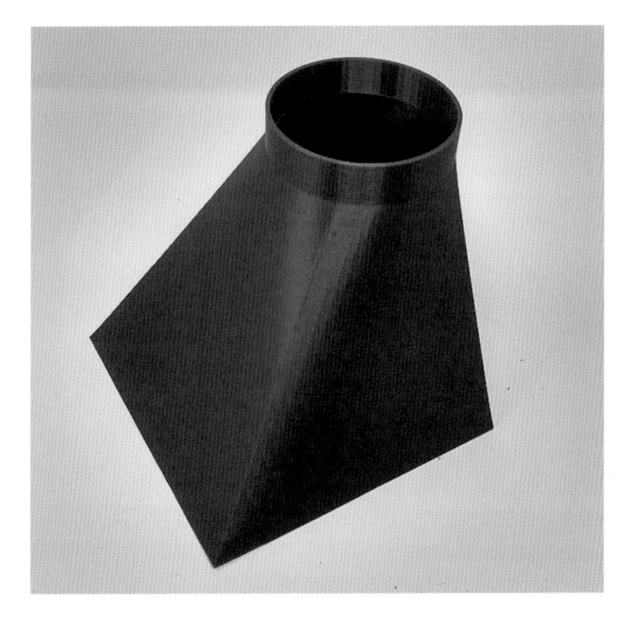

Printed dust collector hood

Clamps

Like the hinged box, the C-clamp is another example of a project that prints separate items at the same time. The C-clamp uses a ⅜-inch bolt with a cap and swivel on the end. The cap and swivel are printed together with support material used to hold up the swivel, preventing the two pieces from joining together.

Print the C-clamp using the following settings:

File: www.thingiverse.com/thing:3535511 (small C-clamp)
www.thingiverse.com/thing:3535578 (basic C-clamp)
Material: PETG
Temperature: 230°C or in accordance with the manufacturer's directions for PETG
Bed temperature: 85°C to 90°C for PETG
Infill: 40%
Support: set supports
Brim: not needed
Raft: not needed

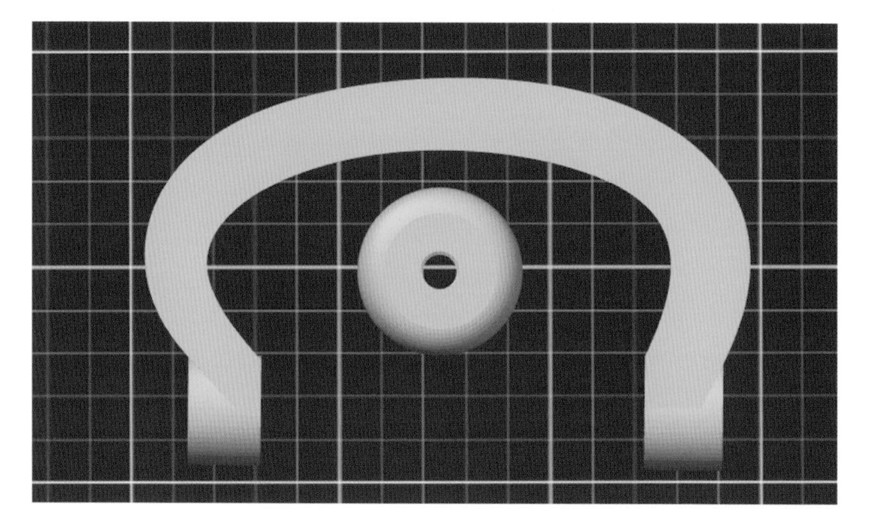

C-clamp parts ready to print

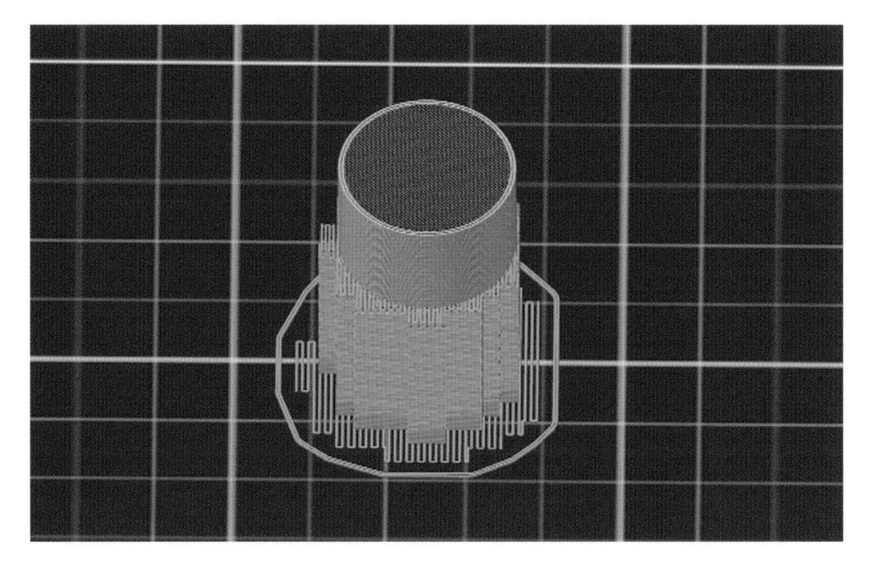

C-clamp rod end with support material

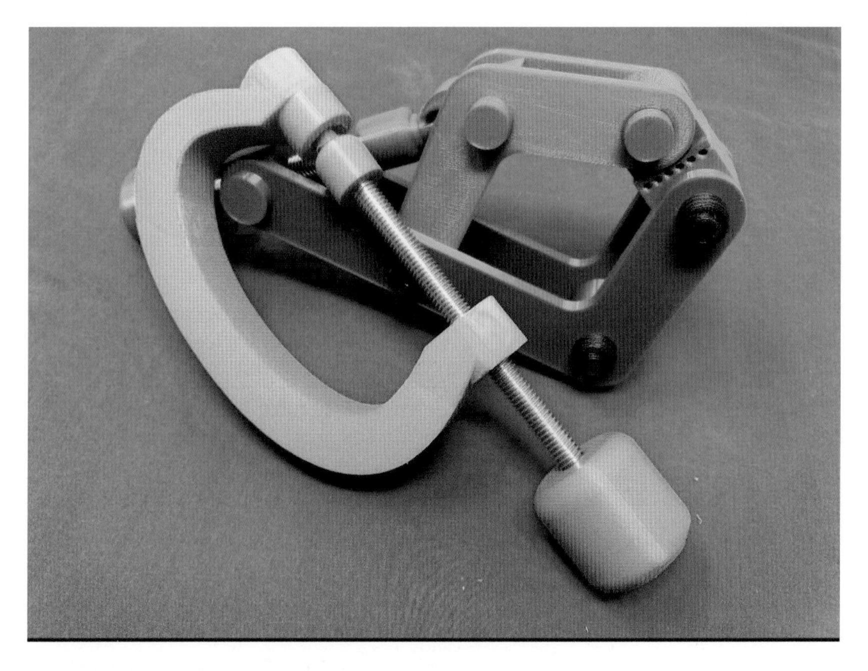

C-clamp and Kant-Twist clamp

Small C-clamps

Kant-Twist Clamp

The Kant-Twist clamp is a clamp design that fixes most of the problems of the C-clamp. The clamp has free floating jaws that adjust to the angle of the work surfaces, and the clamp will not walk off center when pressure is applied against the jaw.

The print design of this clamp uses seventeen printed parts. The parts are all correctly sized for easy printing, and all of the threaded parts are adjusted to compensate for the errors created by the 3D printing process. Each of the parts can be printed without supports.

When you are using fused deposition printing, all parts are slightly overprinted. The tolerances are based on the printer, so different printers will give different results. Your printer may not provide the same level of tolerance, and the size of the threaded parts may need to be adjusted. The threaded shaft must be printed upright to allow the threads to be properly formed.

The printer should have a print volume of at least 8 × 8 × 8 inches (200 × 200 × 200 mm). Print the Kant-Twist clamp using the following settings.

File: www.thingiverse.com/thing:3538259
Material: PETG, carbon fiber nylon, Polycarbonate (PC); higher strength filament is required for this print
Temperature: dependent on filament used and in accordance with the manufacturer's recommendations
Bed temperature: dependent on filament used and the manufacturer's recommendations
Infill: 40%
Support: supports should not be needed
Brim: may be needed on some parts
Raft: may be needed on some parts

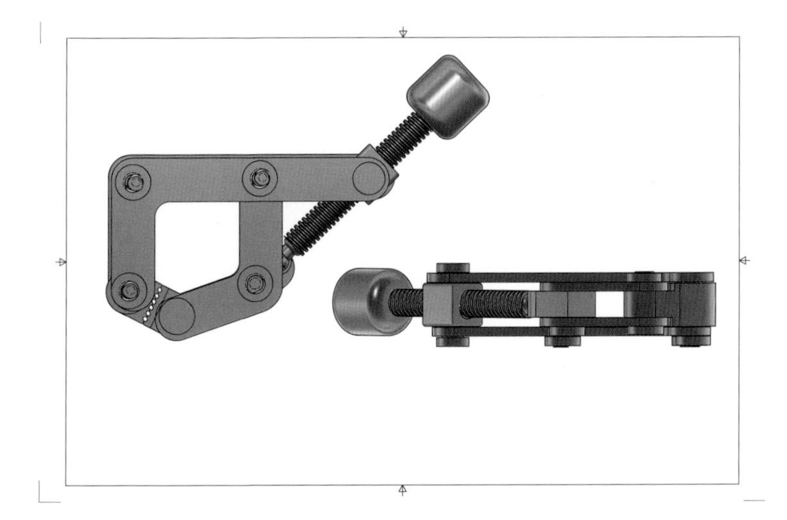

Kant-Twist clamp

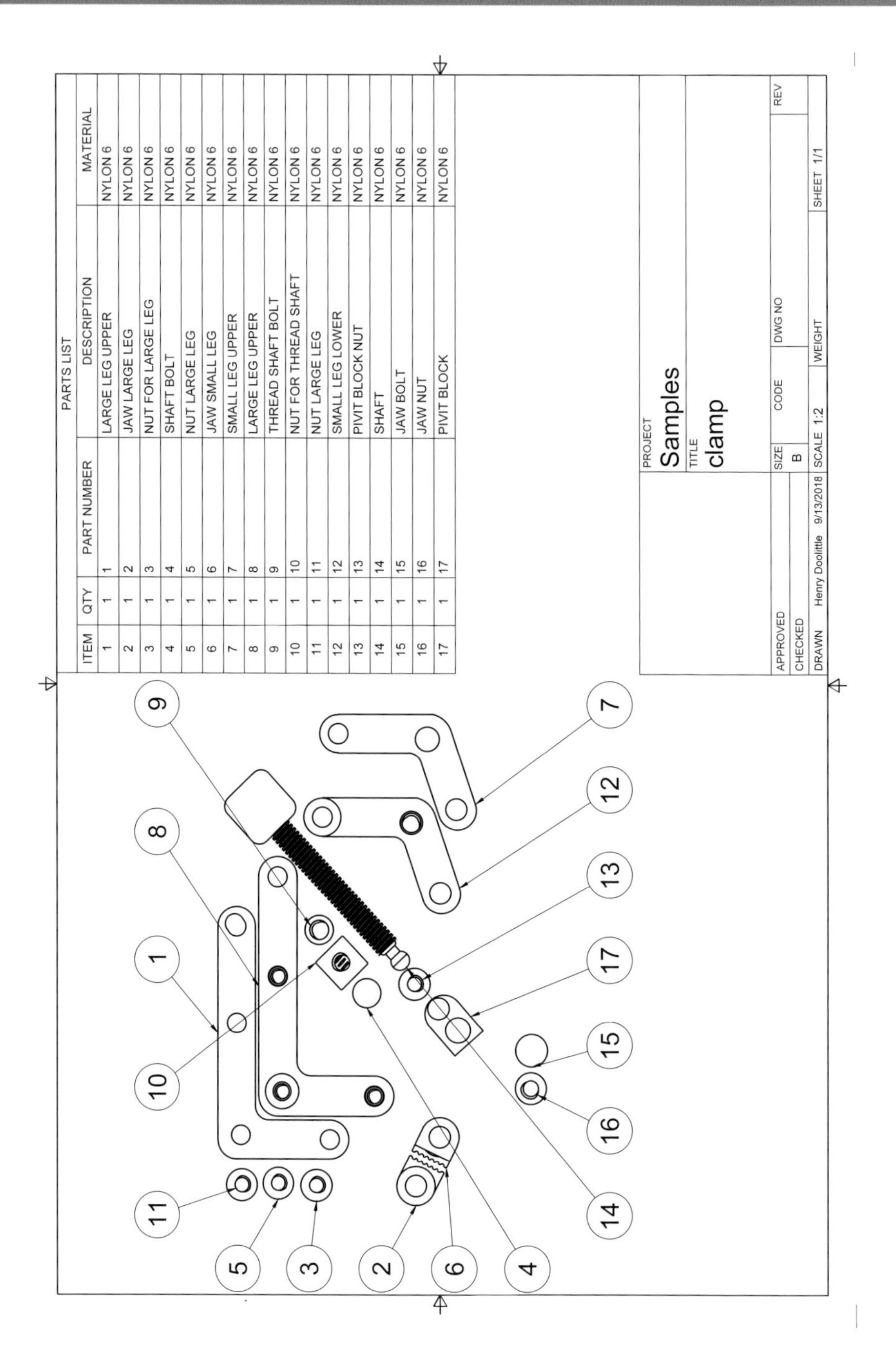

PARTS LIST

ITEM	QTY	PART NUMBER	DESCRIPTION	MATERIAL
1	1	1	LARGE LEG UPPER	NYLON 6
2	1	2	JAW LARGE LEG	NYLON 6
3	1	3	NUT FOR LARGE LEG	NYLON 6
4	1	4	SHAFT BOLT	NYLON 6
5	1	5	NUT LARGE LEG	NYLON 6
6	1	6	JAW SMALL LEG	NYLON 6
7	1	7	SMALL LEG UPPER	NYLON 6
8	1	8	LARGE LEG UPPER	NYLON 6
9	1	9	THREAD SHAFT BOLT	NYLON 6
10	1	10	NUT FOR THREAD SHAFT	NYLON 6
11	1	11	NUT LARGE LEG	NYLON 6
12	1	12	SMALL LEG LOWER	NYLON 6
13	1	13	PIVIT BLOCK NUT	NYLON 6
14	1	14	SHAFT	NYLON 6
15	1	15	JAW BOLT	NYLON 6
16	1	16	JAW NUT	NYLON 6
17	1	17	PIVIT BLOCK	NYLON 6

Parts list for Kant-Twist clamp

This clamp should be printed out of PETG or polycarbonate for strength. Nylon or nylon with carbon fibers would also work. The design attempts to keep the stresses on the arms to the extent possible.

The first step is to print the four arms (parts 1, 7, 8, and 12). The arms should be printed with 50% infill in a crosshatch pattern, face up on the print bed. The print will take ten to fourteen hours to print and use about 100 grams of filament, depending on the printer used and the nozzle size.

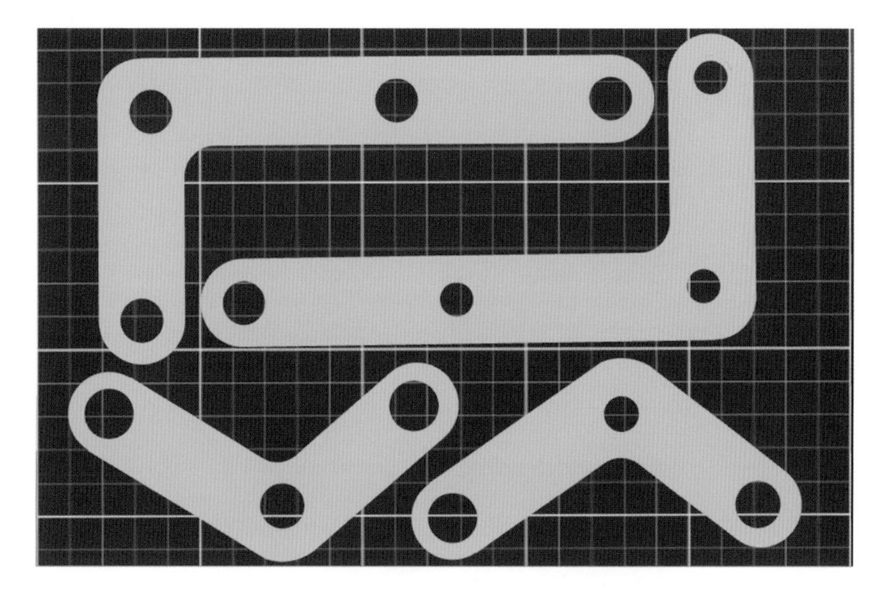

Large and small arms for the Kant-Twist clamp laid out on the print bed

Next print out the nuts, bolts, and blocks, all the remaining parts. These parts can be printed out with 30% infill. The print will take five to six hours and use about 40 grams of filament. Support material is not required, but parts should be placed on the print bed in the orientation shown below.

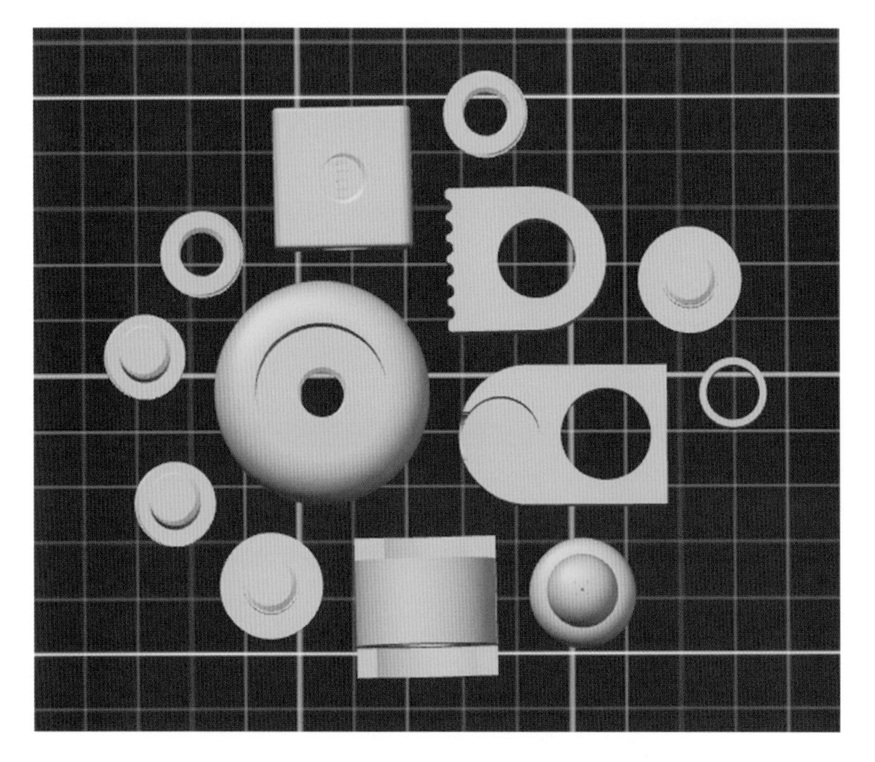

Remaining parts for the Kant-Twist clamp, ready to print

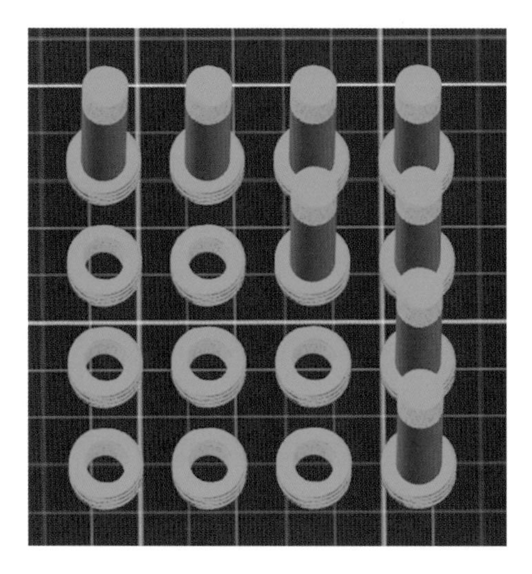

3D printed ⅜-inch nuts and bolts for Kant-Twist clamp

All threaded parts can be purchased from the local hardware store or they can be printed on the 3D printer. The shaft is set up for a ⅜ × 5-inch bolt. Because of the print time and potential for failure, a steel bolt is preferred, but the other bolts work very well when printed on a 3D printer. A ⅜-inch bolt printed with solid fill using carbon reinforced nylon can be as strong as aluminum under the right printing conditions.

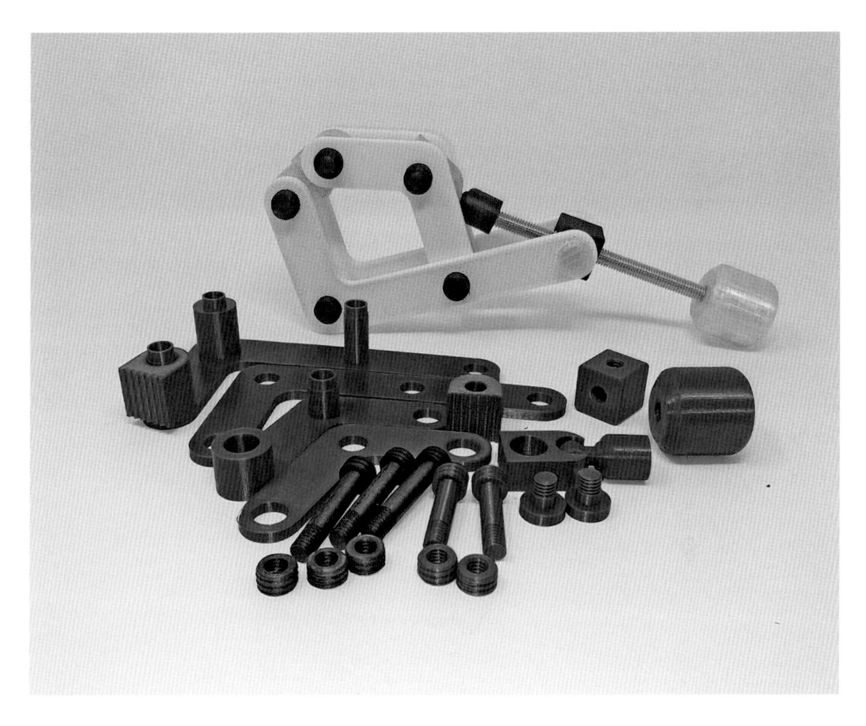

Kant-Twist clamp assembled and with parts separated

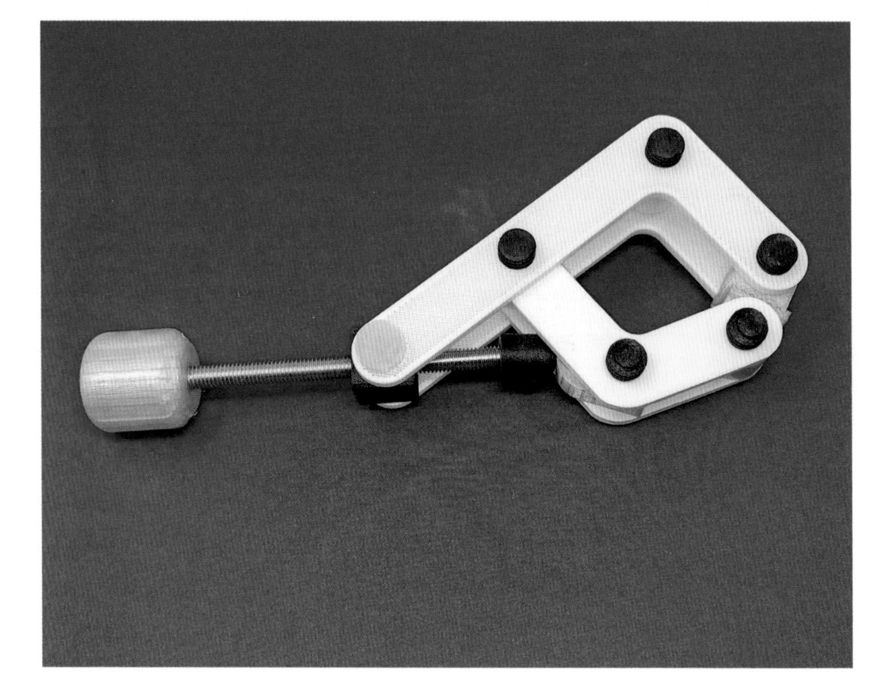

Kant-Twist clamp assembled

Chapter 11

Finish

In theory there is no difference between
theory and practice while in practice there is.

—Benjamin Brewster,
Yale Literary Magazine, February 1882

Do, or do not. There is no try.

—Yoda, *The Empire Strikes Back*

A 3D printer is another shop tool, nothing more. When it works, it works great. When it doesn't work, it's a royal pain. Either way, it is the future. You got this far, and you have seen some of what a 3D printer can do. Now it's time to go out and see how a 3D printer fits in your shop.

You can print custom drawer pulls that match the customer's existing furniture. Prints can be made for custom castings such as specialty hinges for a jewelry box. The possibilities are endless. The design process for simple parts is not complicated. Most parts are built from cubes, cylinders, and spheres. By combining these three shapes, by either joining or cutting them, you can make most of the tools you need.

This book has presented just a small sample of what can be done using a 3D printer. This versatile device can be used to make the final product, parts of the final product, or tools to make the product.

Not all designs are going to work. Not all prints are going to work. But there will be prints and filaments that exceed expectations.

The important thing is to go out and look on the internet, get some free CAD programs, and try them out. What you design and what you make depend only on your imagination.

"Desiderata"

Go placidly amid the noise and the haste and remember what peace there may be in silence. As far as possible, without surrender, be on good terms with all persons.

Speak your truth quietly and clearly; and listen to others, even to the dull and the ignorant; they too have their story.

Avoid loud and aggressive persons; they are vexatious to the spirit. If you compare yourself with others, you may become vain or bitter, for always there will be greater and lesser persons than yourself.

Enjoy your achievements as well as your plans. Keep interested in your own career, however humble; it is a real possession in the changing fortunes of time.

Exercise caution in your business affairs, for the world is full of trickery. But let this not blind you to what virtue there is; many persons strive for high ideals, and everywhere life is full of heroism.

Be yourself. Especially do not feign affection. Neither be cynical about love; for in the face of all aridity and disenchantment it is as perennial as the grass.

Take kindly the counsel of the years, gracefully surrendering the things of youth.

Nurture strength of spirit to shield you in sudden misfortune. But do not distress yourself with dark imaginings. Many fears are born of fatigue and loneliness.

Beyond a wholesome discipline, be gentle with yourself. You are a child of the universe no less than the trees and the stars; you have a right to be here. And whether or not it is clear to you, no doubt the universe is unfolding as it should.

Therefore, be at peace with God, whatever you conceive Him to be. And whatever your labors and aspirations, in the noisy confusion of life, keep peace in your soul. With all its sham, drudgery and broken dreams, it is still a beautiful world.
Be cheerful. Strive to be happy.

—a poem by Max Ehrmann, 1948

Index

About the Author

Henry Doolittle worked in the nuclear industry for forty-four years as a mechanical engineer. About twenty years ago he bought a small lathe and took up woodturning as a hobby. The lathe got him started in woodworking. About six years ago he picked up a CNC router and started making jigs and fixtures, which in turn led to his interest in using 3D printers in the craftsman's shop. Using the 3D printer, he has been able to print his own woodworking tools, a process which this book details. ***3D Printers for Woodworkers*** is Doolittle's first book.